Praise for Rock, Paper, Scissors

'*Rock, Paper, Scissors* brings the evolution of cooperation to everyone with a succinct summary of how these exciting ideas change the way we look at the world and the way we think.'

– *John R. Hauser, Kirin Professor of Marketing, MIT Sloan School of Management*

'This is a super account of the most important unsolved problem in all of science: how did cooperative behaviour evolve, enabling complex human societies to arise and persist. Effective action to address issues such as climate change, biodiversity loss, or feeding tomorrow's still-growing population depends on still better understanding of this problem. Read the book!'

– *Lord Robert May, Zoology Department, Oxford University*

'In its sixty years of evolving, game theory has emerged from a mathematical phase, to a pedagogical phase, to a level of development that can be understood and useful to the intelligent reader. Len Fisher's book clearly outlines the uses of game theory in everyday life in general, and in encouraging cooperative behaviour in particular. A tour de force of exposition, with many amusing and enlightening vignettes of the application of game theory to real-world interactions in the home, amongst friends, in business, and in international relations. A great introduc   or the intelligent reade

– *Professor Rc Management*

ROCK, *Paper,* SCISSORS

ROCK, *Paper,* SCISSORS

Game Theory in Everyday Life

LEN FISHER, *Ph.D.*

HAY HOUSE

Australia • Canada • Hong Kong • India
South Africa • United Kingdom • United States

First published by Basic Books, a Member of the Perseus Book Group

First published and distributed in the United Kingdom by:
Hay House UK Ltd, 292B Kensal Rd, London W10 5BE. Tel.: (44) 20 8962 1230;
Fax: (44) 20 8962 1239. www.hayhouse.co.uk

Published and distributed in Australia by:
Hay House Australia Ltd, 18/36 Ralph St, Alexandria NSW 2015. Tel.: (61) 2 9669 4299;
Fax: (61) 2 9669 4144. www.hayhouse.com.au

First published and distributed in the United Kingdom by:
Hay House UK Ltd, 292B Kensal Rd, London W10 5BE. Tel.: (44) 20 8962 1230;
Fax: (44) 20 8962 1239. www.hayhouse.co.uk

Published and distributed in the Republic of South Africa by:
Hay House SA (Pty), Ltd, PO Box 990, Witkoppen 2068. Tel./Fax: (27) 11 467 8904.
www.hayhouse.co.za

Published and distributed in India by:
Hay House Publishers India, Muskaan Complex, Plot No.3, B-2, Vasant Kunj, New Delhi –
110 070. Tel.: (91) 11 4176 1620; Fax: (91) 11 4176 1630. www.hayhouse.co.in

A catalogue record for this book is available from the British Library.

ISBN 978-1-8485-0202-4

Printed and bound in Great Britain by CPI Bookmarque, Croydon, CR0 4TD.

To Wendella, who coped and encouraged magnificently throughout the development of this book, and Phil Vardy, whose constructive criticism of my first efforts was instrumental in setting me on the right track.

Contents

Acknowledgments

MANY PEOPLE HAVE HELPED ME WITH THIS PROJECT. My wife, Wendella, has been a constant source of inspiration, support, and ideas from the very start, and she has spent a great deal of time and effort in editing draft chapters and suggesting useful ways to improve them. Professor Bob Marks, from the Australian Graduate School of Management, has kindly read the whole manuscript through and offered comments and suggestions from the point of view of a professional game theorist. Dr. Adrian Flitney has done likewise for the section on quantum game theory. Many other people have proposed ideas and have read and commented on draft chapters, including (in alphabetical order) Simon Blomberg, Kathie Davies, Lloyd Davies, John Earp, David Fisher, Diana Fisher, Keith Harrison-Broninski, Virginia King, Alan Lane, John Link, Emma Mitchell, Sue Nancholas, Geoff Oldman, Clive Perrett, Harry Rothman, Al Sharp, Phil Vardy, and my whole 2007 science history tour group. Finally, I must thank my agent, Barbara Levy, who encouraged me to pursue this idea and who has never lost faith in its importance and relevance, and my New York editor, Amanda Moon, whose detailed and perceptive comments have resulted in many improvements to the original manuscript. If I have inadvertently omitted anyone's name from this list, I can only apologise and hope that they will at least enjoy a drink with me the next time that I see them.

Introduction

A FRIEND CALLED ME RECENTLY with the news that a group of scientists had just published a study on how teaspoons gradually disappear from the communal areas of offices. 'Game theory!' he screamed triumphantly. I thanked him profusely, and added yet another example to my already thick file.

Game theory is all around us. Despite its name, it is not just about games – it is about the strategies that we use every day in our interactions with other people. My friends have been sending me examples from newspaper stories and their own personal experience ever since I announced my intention to write a book about it. I wanted to find out whether its surprising new insights could help us develop fresh strategies for cooperation, and to try them out for myself in environments that ranged from the polite confines of an English dinner party to baseball games, crowded pavements, shopping centres, congested Indian roads, and Australian outback pubs.

Game theory tells us what is going on behind the confrontations, broken promises, and just plain cheating that we so often see in domestic quarrels, neighbourhood arguments, industrial disputes, and celebrity divorce cases. It also gives guidance to the best strategies to use in situations of competition and conflict, which is why big business and the military have taken to it like ducks to water since it was invented in the late 1940s. It provides businessmen with strategies to get the better of their competitors, and guides Western military thinking to an alarming extent. Professional game theorists have often had a foot in both camps. To give just one example, all five game theorists who have won Nobel Prizes in economics have been employed as advisors to the Pentagon at some stage in their careers.

But there is another side to game theory – a side that concerns cooperation rather than confrontation, collaboration rather than competition. Biologists have used it to help understand how cooperation evolves in nature in the face of 'survival of the fittest'. Sociologists, psychologists, and political scientists are using it to understand why we have such problems in cooperating, despite the fact that we need cooperation as never before if we are to resolve important and worrying problems like global warming, resource depletion, pollution, terrorism, and war. I wanted to see whether it could be used in everyday situations and to find out whether the lessons learned might be helpful in resolving larger-scale problems. At the least, I thought, I might discover some clues as to how we as individuals could help to resolve such problems.

Game theorists have discovered an amazing link between all of these problems – a hidden barrier to cooperation that threatens to produce untold damage unless we learn to do something about it, fast. The barrier presents us with a catch-22 logical trap that is a constant, if often unrecognised, presence in family arguments, neighbourhood disputes, and day-to-day social interactions, as well as in the global issues that we now face. It even accounts for the way that spoons mysteriously disappear from the communal areas of offices.

The scientists who studied the problem, who were otherwise perfectly sane and respectable Australian medical epidemiologists, had a lot of fun dreaming up unlikely explanations. One was that the spoons had escaped to a planet entirely populated by spoon life-forms, there to live an idyllic existence in which they were not being dunked head-down in cups of hot tea or coffee. Another was resistentialism – the belief that inanimate objects have a natural antipathy toward humans and are forever trying to frustrate us, in this case by hiding when they are most wanted, in the manner of single socks in a washing machine.

The serious explanation, though, was that this was an example of the Tragedy of the Commons – a scenario that was brought to public attention by the Californian ecologist and game theorist Garrett Hardin in a 1968 essay, although philosophers have been worrying about it since the time of Aristotle. Hardin illustrated it with the parable of a group of herders each grazing his own animals on common land, with one herder thinking about adding an extra animal to his herd. An extra animal will yield a tidy profit, and the overall grazing capacity of the land will only be slightly diminished, so it seems perfectly logical for the herder to add an extra animal. The tragedy comes when all the other herders think the same way. They all add extra animals, the land becomes overgrazed, and soon there is no pasture left.

The scientists applied the same argument to teaspoons: 'teaspoon users (consciously or otherwise) make decisions that their own utility [i.e., the benefit to themselves] is improved by removing a teaspoon for personal use, whereas everyone else's utility is reduced by only a fraction per head ("after all, there are plenty more spoons . . ."). As more and more teaspoon users make the same decision, the teaspoon commons is eventually destroyed.'

It sounds funny when applied to teaspoons, but if you replace the word *teaspoon* with *land*, *oil*, *fish*, *forest*, or the name of any other common resource, you will soon see that some very serious global problems have their origins in this vicious circle of logic, which can make its unwelcome presence felt whenever profit goes to an individual person or group of people but costs are shared by the community as a whole.

The Tragedy of the Commons exerts its destructive power whenever some of us cooperate for mutual benefit but others see that they could do better for themselves by breaking the cooperation (in game theory parlance, *defection* or *cheating*). So they can, until everyone else starts thinking in the same way, when the cooperation collapses and everyone ends up worse off. Through following the logic of self-interest, they have somehow landed everyone in a position where self-interest is the last thing that is being served.

This intractable logical paradox links the collapse of the Newfoundland cod fisheries, the ruinous civil war in Sudan, China's massive expansion in fossil fuel-driven power stations, and the tendency of many Americans to drive wasteful gas-guzzling cars. It underlies spam on the Internet, burglary, queue jumping, and many traffic accidents. It was probably the logic that led to the felling of the last tree on Easter Island. It is certainly the logic that leads people to dump their household waste on an empty site instead of disposing of it properly, and to exaggerate insurance claims or "forget" to declare income on tax forms. It is also the logic that governments use when they refuse to sign international agreements like the Kyoto Protocol.

Most importantly, it is the logic of escalation. In the words of the great 1970s protest song:

> *Everybody's crying peace on earth,*
> *Just as soon as we win this war.*

When both sides use the same logic, however, there is never going to be any peace in this world.

We could avoid the Tragedy of the Commons if we were to change our behaviour and become more moral or more altruistic, caring for our neighbours at least as much as we do for ourselves. It would be great if this were to happen, but the reality is that we are not all Mother Teresas, and we had better face the fact that we often cooperate only when we can see something in it for ourselves. This applies to nations as much as it does to individuals; the author of the influential 2006 'Stern Review on the Economics of Climate Change' made the point, for example, that nations would only cooperate to solve the problem if they could see some direct, short-term economic benefit to themselves.

Game theory makes no moral judgments about such attitudes. It simply accepts the fact that self-interest is one of our primary motivations and judges different strategies according to how they serve that interest. The paradoxes and problems come in when a strategy of cooperation would lead to the best outcome for all concerned but each party is tempted to try for a better outcome for itself, only to become trapped by its own greed in an inferior situation, like a lobster caught in a pot.

There's not much point in criticising the greed, although it would certainly help if people (and nations) were content to accept no more than their fair share of the world's resources. What is more important is to understand the trap, which is the first step in finding ways to avoid it or escape from it, and reach co-operative solutions to problems instead.

The trap has been with us since time immemorial. Examples can be found in the Bible, the Koran, and many ancient texts, as well as in history books, the plots of novels and operas, and many modern news stories. Its true nature was not understood until the late 1940s, though, when the advent of game theory permitted the Nobel Prize-winning mathematician John Nash (the schizophrenic anti-hero of the film *A Beautiful Mind*) to reveal its inner workings.

Those inner workings are the central theme of this book. They catch us in a series of *social dilemmas* to which game theorists have given evocative names. One is the Tragedy of the Commons. Another is the famous Prisoner's Dilemma, which is exemplified by the U.S. practice of plea bargaining, and which is the subject of chapter 1. Others are the game of Chicken (which nearly led to world catastrophe when Kennedy and Khrushchev played it during the Cuban Missile Crisis), the Volunteer's Dilemma (en-capsulated by the word *mamihlapinatapai*, of the Yag·n language of Tierra del Fuego, which means 'looking at each other with each hoping that the other will do something that you both want to have done but which neither of you wants to do themselves'), and the Battle of the Sexes (in which a couple wants to go out together rather than separately, but he wants to go to a baseball game while she wants to go to the opera).

Cooperation would lead to the best overall outcome in all of these cases, but Nash's trap (which is now called a *Nash equilib-rium*) draws us by the logic of our own self-interest into a situation in which at least one of the parties fares worse but from which they can't escape without faring worse still. (That is why it is such an effective trap.) If we are to learn to cooperate more effectively, we need to find ways to avoid or escape from the trap. Game theory identified the problem. Can game theory provide any clues that might help us to resolve it? The answer is yes.

Some of those clues have come from studies of the evolution of cooperation in nature. Others have come from a close examination of the strategies that we have traditionally used in our efforts to win and maintain cooperation. Promising strategies for cooperation that have emerged include variations on the I Cut and You Choose theme, new methods of cooperative bargaining (including an amazing application of quantum mechanics), eliciting trust by ostentatiously limiting your own options to cheat or defect, and changing the reward structure to remove the temptation to break cooperative agreements.

Some of the most significant clues have come from computer simulations in which different strategies were pitted against each other to find out which would succeed and which would fall by the wayside. The initial results appeared in Robert Axelrod's book *The Evolution of Cooperation*, which was published in 1984 by Basic Books. According to a later foreword by the biologist Richard Dawkins, 'the world's leaders should all be locked up with this book and not released until they have read it'. Judging by the history of the last twenty years, few world leaders have taken the opportunity to look at the problem of cooperation in such a new and constructive way.

The crunch point is the tit-for-tat strategy (and subsequently discovered variants), which can lead to the escalation of conflict, but can also lead to you-scratch-my-back-and-I'll-scratch-yours cooperation, both in nature and in our own society. It can be a very tight question as to which will emerge, with just a small change in circumstances making a vast difference to the outcome, as happens in boom-and-bust economic cycles and in the expansion and contraction of animal populations. Mathematicians call the critical point a *bifurcation point*, with the prospect of two very different futures depending on which path is followed. The problem of cooperation is often the problem of finding a strategy that will tilt the balance of tit for tat toward a cooperative, backscratching future rather than one of escalating conflict.

Recent studies have offered some tantalising hints as to how this might be achieved. That's not to say that game theory offers a panacea – that would be a ridiculous claim – but it has certainly provided new insights into the way cooperation evolves and suggested new strategies and new twists to the old strategies. In this book I describe my efforts to understand these strategies and to try them out for myself in everyday situations. My aim was to assemble a toolkit of potential strategies for cooperation, in the same way that I have built up a toolkit of techniques for tackling scientific problems during my life as a scientist. I have had a lot of fun during that life but never so much as when I was performing these experiments on cooperation. The results were sometimes hilarious, sometimes alarming, but invariably enlightening in providing lessons about just what it takes to get people to cooperate – and to keep cooperating.

Finally, I should emphasise that I am not a professional game theorist but a scientist and concerned human being searching for answers to some of our most pressing social questions. Game theory illuminates these questions from a perspective with which many people will be unfamiliar and I wanted to find out just how relevant its answers might be to the problems of real life. I hope that you enjoy sharing my journey of discovery.

The Organisation of This Book

The book begins with a chapter on the basic nature of the Nash equilibrium, showing how it leads to the famous Prisoner's Dilemma, which underlies many of our most serious problems (including the Tragedy of the Commons). This is followed by a chapter on ways to divide resources fairly using strategies such as I Cut and You Choose. My conclusion in these two chapters is that we can't rely on external authorities or on our own sense of fairness to produce lasting cooperation, and that we must look more deeply at how we can use our own self-interest to make the cooperation self-enforcing.

In chapter 3 (a key reference chapter) I use game theory to examine how different social dilemmas actually arise. This is followed by a series of chapters on strategies for cooperation that include a remarkable variant on the childhood Rock, Paper, Scissors game, new methods of cooperative bargaining, methods of eliciting trust, and the use of tit-for-tat strategies. I show how such strategies emerge in nature, and investigate how we might be able to use them to promote cooperation rather than confrontation in our own society. I then investigate how we might avoid social dilemmas by changing the game itself, either by introducing new players or by an amazing application of quantum theory. Finally, I review the strategies for cooperation that I have uncovered and present my personal top ten list of tips for effective strategies in different situations. If you want to see how it all pans out, feel free to take a glance at this chapter first.

As with my previous books, there are extensive notes at the back that contain anecdotes, references, and expanded discussions of some points that could not be comfortably fitted into the main chapters. These are designed to be read independently and can be dipped into just for fun. Some readers of my previous books have even written to say that this is where they start!

A Note of Explanation

As I pursued my investigation I became painfully aware that almost any paragraph could have been expanded into a major article, if not a full book. In order to keep this book shorter than the *Encyclopaedia Britannica*, I have minimised or omitted discussions of many complicating factors. If the reader is sufficiently stimulated to want to pursue these further, they can be found in any standard textbook on game theory. The main ones are:

- **Nash's Trap.** Professional game theorists may not much like my describing the Nash equilibrium in this way, because it implies that the equilibrium always leads to a bad outcome. I am sticking with it, though, because this book is about bad outcomes and how to get out of them. The reader should be aware, though, that the trap actually comes in three varieties: tender, tough, and terrible. The tender version is one in which we are trapped into the same set of strategies that we would have come up with if we had agreed to cooperate for mutual benefit. This sort of trap doesn't get much attention in this book, although it does make a walk-on appearance in chapters 5 and 6. Most of the book is concerned with the tough and terrible traps that land us in social dilemmas.

- **N-person Situations.** Cooperation can be between two individuals (or groups of individuals), or it can involve many individuals or groups. I have kept my examples mainly to the former, with an occasional bold excursion to the more complicated case.

- **Perfect and Imperfect Information.** Game theorists distinguish between the two situations. So do I, but without saying so. Sometimes we have a clear knowledge of someone else's past actions. Sometimes we have to use what information we have to make an educated guess. It will usually be obvious from the context which of these two situations I am describing.

- **Simultaneous or Sequential Strategic Decisions.** We can make strategic decisions without knowing what the strategy of the other party is (game theorists call this simultaneous), or we can make them after the other party has made and acted on theirs and we know what they have done (sequential). It will again be obvious from the context which sort of situation I am talking about.

- **Rationality.** There is a lot of discussion among game theorists and others about just what it means to be rational. Maybe the sort of logic that leads to the Tragedy of the Commons and other social dilemmas isn't really so rational. Sometimes, also, it turns out that the most rational thing that we can do is to appear to be irrational! All of these points will come up in the course of this book.

LEN FISHER
Bradford-on-Avon, U.K.
and Blackheath, Australia
May 2008

1

Trapped in the Matrix

THE HIDDEN LOGICAL TRAP that John Nash discovered pervades our lives. It leads us into a devastating series of social dilemmas – the game theorist's rather insipid term for situations like the Tragedy of the Commons, in which cooperation would produce the best overall outcome but individuals can be tempted by the logic of self-interest to cheat on the cooperation. When both sides cheat, however, the results can be catastrophic, as the characters in Puccini's opera *Tosca* discover when they are caught in the situation that game theorists now call the Prisoner's Dilemma.

Tosca, the heroine of the plot, is faced with an unenviable choice. Her lover, Cavaradossi, has been condemned to death by the corrupt police chief Scarpia. Tosca is left alone with Scarpia, who thinks that he is on to a good thing when he offers to have the firing squad use blanks if Tosca will let him have his wicked way with her. What should Tosca do? She spies a knife on the table and figures out that she can win both ways by agreeing to Scarpia's proposal, but actually stabbing him when he comes close. Unfortunately for her, Scarpia has already worked out that *he* can win both ways by not really telling the firing squad to use blanks. He dies, Cavaradossi dies, and when Tosca finds out what has happened, she flings herself off a castle parapet and dies too. Everyone is a loser, as is often the way with opera.

Everyone is a loser in real life as well when caught in what game theorists call the Prisoner's Dilemma, after an example used by Princeton University mathematician Albert Tucker to illustrate the problem to a group of psychologists in the early 1950s.

The story has since appeared in various incarnations. In one of them, two thieves (let's call them Bernard and Frank, after two of the conspirators in the Watergate scandal) have been caught by the police, but the prosecutor has enough evidence to put them behind bars for only two years, on a charge of carrying a concealed weapon, rather than the maximum penalty of ten years that they would get for burglary. So long as they both plead not guilty, they will both get only two years, but the prosecutor has a persuasive argument to get them to change their pleas.

He first approaches Bernard in his cell and points out that if Frank pleads guilty but Bernard doesn't, Frank will receive a reduced sentence of four years for pleading guilty, but Bernard will get the maximum ten years. So Bernard's best bet, if he believes that Frank will plead guilty, is to plead guilty as well and receive four years rather than ten. 'Furthermore,' says the prosecutor, 'I can offer you a deal that if you plead guilty and Frank doesn't, you can go free for turning over state's evidence!'

No matter what Frank does, it seems that Bernard will always do better for himself by pleading guilty. The logic seems irrefutable – and it is. The trouble is that the prosecutor has made the same offer to Frank, who has come to the same conclusion.

So they both plead guilty – and they both end up sentenced to four years, rather than the two years they would have received if they had both kept their mouths shut.

If you think that this little story has uncomfortably close parallels with the U.S. legal practice of plea bargaining, you are dead right. This is why the practice is outlawed in many countries. The logical paradox illustrated by the story affects us in many situations, from divorce to war – so many, in fact, that it has been proposed as the basic problem of sociology, since our efforts to live together in a cooperative and harmonious way are so often undermined by it.

It certainly undermined my young brother and me when we stole a cake that our mother had made and gorged ourselves with it. We could have escaped punishment, and the dog might have received the blame, if we had both kept our mouths shut, but I thought it would be less risky to lay the blame on my brother. He had the same idea, however, and we were both confined to our rooms with our aching stomachs and backsides.

The insidious logic of the Prisoner's Dilemma caught us out again in our late teens when each of us developed an interest in the same girl, whose family had just moved to the neighbourhood and joined our local church. We weren't the only boys whose interest was sparked by the attractive new arrival, but our efforts to snare her in our adolescent nets were doomed to failure as soon as each of us tried to win the day by telling her undermining stories about the other. It wasn't long before we saw her going out with another boy altogether.

The Prisoner's Dilemma is always with us. Another nice example from the United Kingdom concerns price-fixing by supermarkets after the 2002–2003 foot-and-mouth disease outbreak that led to many dairy cattle being slaughtered. Four large supermarket chains raised the price of milk, butter, and cheese, saying that they were paying more to farmers to help keep them in business. They weren't – two at least were just pocketing the extra profits. These owned up to it after they were charged with collusion by the Office of Fair Trading, and they pointed the finger at the other two (who denied price-fixing) in return for a much reduced fine compared to what the others will get if they are found guilty.

Yet another example comes from the history of the Dead Sea Scrolls, which were discovered at the Qumran cave site near the northwest corner of the Dead Sea. After the first scrolls were found, the Bedouin shepherds in the area discovered that archae-ologists were willing to pay high prices for them, and the shep-herds began to look for more, finding some in a rather dilapidated condition. They had also discovered that the archaeologists were willing to pay piece rates for the fragments, so they began to tear up intact scrolls in order to offer them progressively as separate pieces! The archaeologists could have escaped from the situation only by paying disproportionate sums for larger pieces. Other-wise, the shepherds could only lose by offering larger pieces. Together, they were trapped in a Prisoner's Dilemma, to the detri-ment of biblical scholarship and culture.

The Prisoner's Dilemma presents us with a logical conun-drum that lies at the heart of many of the world's most seri-ous problems. The arms race that began in the 1950s is a good example. Cooperation to limit arms production and save the money for more constructive purposes would have benefited everyone, but no nation could benefit from unilaterally disarm-ing so long as other countries continued to build up stocks of nuclear weapons. More recently, our efforts to resolve the threat of global warming are being hampered by the same paradoxical logic, because many polluting nations feel that there is little incentive to control their carbon emissions so long as other nations continue to pollute.

The physical sciences can't resolve such problems for us in the long term; the best that they can hope to do is to ameliorate them in the short term. To do any better, we need to develop a deeper understanding of ourselves. This was one reason why I took some time off from science to study philosophy, in the hope of finding some answers. What I found, though, brought me right back to science. I soon discovered that the whole field of ethics, which is concerned with the principles that we should live by to create a stable and just society, comes down to the story of historical attempts to get around the problems exemplified by the Prisoner's and other social dilemmas, which have their basis in logic and mathematics. I rather enjoyed delving into the mathematics and the formal logic, but fortunately one needs neither of these to understand where the problems come from and how they affect us.

· The great breakthrough in understanding social dilemmas came in 1949, when John Nash discovered that all of them arise from the same basic logical trap. Nash is now familiar to many people as the anti-hero of *A Beautiful Mind*, but the film focused almost exclusively on his mental illness. It gave little indication of what his Nobel Prize-winning discovery actually was or how incredibly important it is to our understanding of the problems of cooperation and what we might be able to do about them.

Nash made his discovery when he was just twenty-one and not yet suffering from the schizophrenia that was to blight much of his life. He is even able to joke about his mental illness, saying in one interview: 'Mathematicians are comparatively sane as a group. It is the people who study logic that are not so sane.'

He had arrived at Princeton University in 1948 to study for a postgraduate degree in mathematics, bearing a laconic one-line recommendation from his previous professor: 'This man is a genius.' He proved his genius within eighteen months by using the recently developed science of game theory first to identify the logical trap (now known as the 'Nash equilibrium') and then to prove a startling proposition – that there is at least one Nash equilibrium lying in wait to trap us in *every* situation of competition or conflict in which the parties are unwilling or unable to communicate.

The idea behind the Nash equilibrium is deceptively simple (see Box 1.1). It is a position in which both sides have selected a strategy and neither side can then independently change its strategy without ending up in a less desirable position. If we're walking towards each other on a narrow pavement, for example, and we both step aside to squeeze past, we'll find ourselves in a Nash equilibrium because if either of us independently changes our mind and steps back, we will come face-to-face again, with the consequent merry little dance that most of us have experienced.

Nash called such a state of affairs an *equilibrium* because it is a point of balance in a social situation, from which neither side can independently escape without loss. Note that word *independently* – it is key to what follows. So long as we act independently, with each of us pursuing our own interests, the Nash equilibrium will continue to trap us in a plethora of social dilemmas. If the two people walking along the narrow pavement both independently decide that they would prefer the side further from the gutter, for example, their attempts to improve their own situation by avoiding being splashed by passing cars will mean that they can't get past each other without one or the other giving way.

➤ BOX 1.1

THE NASH EQUILIBRIUM AND THE PRISONER'S DILEMMA

The game theorist's way of describing a Nash equilibrium is that *if each party has chosen a strategy, and no party can benefit by changing his or her strategy while the other parties keep theirs unchanged, then that set of strategy choices and the corresponding payoffs constitute a Nash equilibrium.* Game theorists use shorthand diagrams to summarise the choices and rewards in the same way that builders use diagrammatic plans of a house to make sure that all of the bits fit together. The possibilities are laid out in a matrix that represents the reality in which the participants are trapped, much as they are in the 1999 science fiction film *The Matrix.* To make the comparison stronger, this representation of reality was devised by the Hungarian-American mathematical genius John von Neumann, the inventor of game theory.

Here are Bernard's and Frank's prison sentences for their various choices in the Prisoner's Dilemma, represented as this sort of matrix:

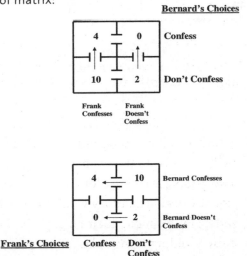

I have drawn little passages between the cells, with arrows pointing in the direction that Frank and Bernard can move to reduce their sentences. These diagrams make it obvious that their logical choice is always to confess, no matter what the other does. Game theorists would say that confess is the dominant strategy, since it is the strategy that leads to the best outcome no matter what the other party does.

Game theorists combine these two diagrams into one, which still contains all the information, but which can be harder to interpret at a glance without practice:

Frank

Confess   Don't Confess

	Confess	Don't Confess
Confess	4 , 4	0 , 10
Don't Confess	10 , 0	2 , 2

(Bernard labels the rows on the left)

This type of diagram makes the *pairs* of outcomes obvious, with Bernard's on the left and Frank's on the right in each cell. It shows, for example, that (0,0) is not an option, because one prisoner can only get off scot-free if the other gets ten years (i.e., their only choices are [0,10] or [10,0]).

If we add little corridors between the cells, with the proviso that Bernard can only move from one choice to another in the vertical direction while Frank can only move horizontally (as is obvious from the earlier diagrams), and we follow their chosen movements by putting the arrows back, it becomes clear why Frank and Bernard are in such a pickle. The cooperative choice (both keeping their mouths shut) is the (2,2) option, but the moment that one or the other tries to do better for himself, the chain

➤

of arrows inexorably takes them to the (4,4) cell, from which they can never escape, because there is no choice arrow leading out of that cell for either of them! This time, I have added their faces to show what they think of their various possible positions:

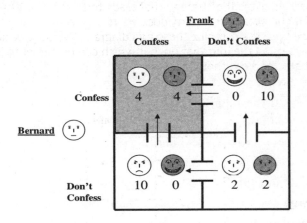

The (4,4) cell represents a Nash equilibrium (drawn here and elsewhere with a grey background) because neither prisoner can independently get out of it without ending up in a worse position. If Bernard chooses not to confess, for example, he will end up in jail for ten years instead of four, and the same thing goes for Frank. Only by making the cooperative, coordinated move of *both* not confessing can they get to the (2,2) option.

The secret to resolving such situations is for the parties to find some way of agreeing to coordinate their actions and for all parties to stick to the agreement. A friend of mine saw a hilarious example of what can happen when these conditions are not fulfilled while he was driving on a mountain road in Italy. A short section of the road had been reduced to one lane. Cars coming from opposite directions were taking turns to pass through it by implicit mutual agreement until two drivers coming from opposite directions each decided to race the other. They came face-to-face in the middle of the narrow stretch, each honking furiously for the other to back up. Neither would budge, and other cars soon crowded in behind them, their horns honking furiously as well. It took the authorities *three days* to clear the resultant traffic jam.

'It served them right,' you might think, 'for being so selfish.' You would be right, but the real problem was that each was acting *independently* in what they perceived to be their own best interest. This is something that we often do and that can land us in devastating Nash traps, as Tosca and Scarpia, and Frank and Bernard, discovered. In one of the shortest scientific papers ever to win its author a Nobel Prize, Nash used a combination of symbolic logic and advanced mathematics to prove the ubiquity of his trap in non-cooperative situations – that is, situations in which the parties are not willing or able to communicate.

Before Nash published his paper, our frequent failure to co-operate with others for mutual benefit was usually thought of in terms of our psychology or our morality, or both. These are obviously important factors, but Nash demonstrated that a deep-seated problem in logic often lies at the heart of such problems, and that it can frequently be the dominant factor. This logical conundrum, baited with the appeal of our own self-interest, repeatedly draws us away from the cooperation that would serve us best and into situations that serve our interests much less.

Just look through any newspaper or celebrity gossip magazine and you'll find examples of the sort of logic that Nash was talking about. Think of two people involved in an acrimonious divorce. It would usually pay both parties to compromise, but so long as one refuses to compromise, it is not worth the other party's while to give way. They become trapped in a Nash equilibrium, so both lose out through the money they have to pay to lawyers and the emotional stress they end up going through.

It is important to emphasise that the parties are trapped in a genuinely paradoxical circle of logic that arises because they are unwilling or unable to communicate and to coordinate their strategies. But there is an escape clause: if the parties can communicate and negotiate, they may be able to break out of the dreadful trap.

Unfortunately, this is not as easy as it sounds. Too often, parties will agree to a negotiated compromise and then one party will break the agreement when it suits them. The problem is that if the cooperative solution (a negotiated agreement) is *not* a Nash equilibrium, one or both sides can generally do better by subsequently changing their strategy. This is a major problem in general, and solving the question of cooperation involves two major challenges: finding some way to reach coordinated agreements and finding some way to make people stick to those agreements. The latter must be sufficiently robust that each side will trust the other to stick to the agreement, and to have that trust justified by results.

This book is about my search for answers to these two major challenges to cooperation, both on a personal level and in the context of the major issues with which we are faced. I discovered that there were three main approaches to meeting the challenges, each favoured by different groups of people and by different cultures. They are:

Changing Our Attitudes: If we came to believe that it was immoral to cheat on cooperation, for example, that would obviously help to resolve many social dilemmas.

Benevolent Authority: Relying on an external authority to enforce cooperation and fair play.

Self-Enforcing Strategies: Developing strategies that carry their own enforcement so there is no incentive to cheat on cooperation once it has been established.

Here I examine all three, arguing that only the third is viable in the long term and that the fresh insights of game theory can help us to devise such strategies in many cases.

Changing Our Attitudes

Philosophers and spiritual leaders have long argued that the road to cooperation is made more difficult by our own greed, selfishness, fear of people who are different from ourselves, and mistrust and ignorance of cultures and beliefs that are different from our own. Can we really expect people to change these attitudes? This was the question I asked a senior Church of England bishop when I debated the question of future cooperation with him in the unlikely setting of the garden of an English country pub.

The event was part of a local cultural festival, and the beer-drinking audience was looking forward to a science-versus-religion confrontation. They must have been disappointed when I agreed with him that the principles of Christian ethics could help to solve the problems of cooperation. 'No one could argue with these principles,' I said, 'and they would certainly work if everyone (or even a sufficient number of us) adopted them. So would the principles advocated by the Dalai Lama: compassion, dialogue, and the "secular ethics" of human values. But what can we do when people *don't* adopt such caring principles and attitudes?'

His answer was that there is little or no hope for a peaceful and cooperative future unless people *do* adopt them. The people in the audience pricked up their ears when I said that I could respect his answer but that there were at least two other answers – one from history and one from science. The one from history is that strong authority, superior force, and divide-and-rule strategies can produce relatively stable societies that can last for long periods of time, albeit at the expense of individual freedoms. The one from the science of game theory is that it is at least sometimes possible to devise strategies for cooperation that do not rely on any of these measures.

'There is one other answer,' I continued. 'We could sit back and wait for evolution to do the job for us. It has solved the problem for species such as ants, bees, and wasps by genetically programming them to cooperate, albeit at the expense of their own individuality. Maybe the human race will also eventually evolve a cooperation gene, and that will solve the problem.'

I could tell that he knew I was pulling his leg, because a broad grin spread over his face. We both knew that it would be ridiculous for us to sit back and rely on nature to help solve the problems of human cooperation. Its solutions can often be drastic, including major changes and even wholesale extinction. But evolution (or a divine source, depending on your point of view) has given us the ability to think problems through for ourselves. Is there some way that we can think this one through?

Benevolent Authority

One answer to the problem of cooperation that has been suggested by philosophers since at least the time of Plato has been to rely on an external authority to see fair play. Plato's particular answer was probably the most impractical of the lot. It was to rely on rule by a set of philosopher-kings (trained by philosophers such as himself, of course). Judging by some of the philosophers I met when I was studying the subject, this would be a surer route to anarchy than most.

Plato's idea was that his philosopher-kings would be benevolent rulers, which is fine in theory but walks bang into the Prisoner's Dilemma in practice. Take King Solomon. Wise he might have been, and benevolent, but he could afford to be benevolent because he had annexed most of his country's wealth for himself. In other words, instead of being a benevolent ruler who distanced himself from the competition for resources and simply oversaw their equitable distribution, he cheated by joining in the competition for those resources. His yearly take of gold alone was around 600,000 troy ounces, which equates to £311 million in today's money. This puts him in the Bill Gates class when taken with his other wealth (including the £39 billion that was left to him to build his famous temple), with the small difference that Solomon's wealth was derived from taxing his people rather than selling things to them.

By joining in the competition for resources, he became part of the problem instead of the key to its solution. That's the issue with relying on authority in general. Authorities can have their own agendas, and these are not always consistent with cooperation and fair play. As soon as they start to follow these agendas, they become a part of the problem instead of the key to its solution.

This can even apply to parents and teachers, the benevolent authorities of our childhood. My own father prided himself on his fairness, but he spent more time and effort on me than he did on my brothers because I happened to perform quite well in exams. His own education had been disrupted, and this was his way of vicariously enjoying the benefits of education.

Let's face it – benevolent authority is largely a myth. We would certainly love to have access to it when we read of bullying in schools, army generals grabbing power in some far-off country, or innocent people being massacred in civil wars. Surely, we think, there must be someone who could act as a powerful independent arbiter to stop these things – a teacher, a big power, or even a world body like the United Nations. But the truth, which screams at you from any newspaper, is that authority needs power, and those with power almost invariably use it to pursue their own interests. Benevolence, however much the powerful might preach it, is the last thing on their minds.

Most autocratic rulers throughout history have used their power to implement their own ends. Philosophers, political theorists, and political activists have tried to get around the problem by placing limits on power, usually by spreading it among members of some small group within the community, or even throughout the community (this is the principle of democratic and communist societies alike). This sounds like a good idea in theory, but in practice the same problems are still there, albeit in different forms, which means that those of us who live in democracies shouldn't be too complacent. We may not have absolute rulers, but we frequently have a majority or majorities that are in a position to suppress the interests of minorities, and often do. Small groups of people can also carry disproportionate power, especially when wealth is involved. Individuals may feel that they are represented, but many analyses of voting systems (see p. 49) have shown that equal representation can be as much a myth as benevolent authority. Elected representatives themselves often kowtow to vested interests and have even been known to take bribes. Certainly legal and judicial systems can play the role of an independent authority, but the law can also be a tool used by those in power. In the immortal words of Charles Dickens' Mr. Bumble, it can even be an ass when judges rely on the letter of the law instead of its commonsense interpretation.

The law can also be powerless in many commonplace situations. If someone pushes into a line of traffic or fails to do his fair share of the work in a communal enterprise, it's not much use shouting for the law. It's not much use shouting for it in serious international situations either. Sometimes it can help to maintain an unsteady peace, as it has in the divided country of Cyprus, for example, but more often it is ineffectual (just think of how many appeals to abide by the United Nations Core International Human Rights Treaties are totally ignored by the offending country) or it becomes a tool of the more powerful side (in the case of the UN, mainly those that hold the power of veto). How else, though, are we to enforce cooperative agreements? Is there another way? Game theory suggests that there is.

Self-Enforcing Strategies

The game theory approach is to avoid the need for an external authority by using the Nash equilibrium as a self-enforcing mechanism to ensure that there is no incentive to cheat on co-operation. This is easily achieved if the cooperative solution is a Nash equilibrium (as is the case in my example of two people approaching each other along a narrow pavement), because in this case it would not pay either party to change their sidestepping strategy. It is much more difficult when the cooperative solution is *not* a Nash equilibrium, because we are then (by definition) in a social dilemma, and there is always a temptation for one or both parties to cheat in the hope of doing better by breaking the cooperative agreement (which, of course, they can until the other party decides to cheat as well, and they both lose out).

In the rest of this book I explore ways in which this might be achieved, in both everyday situations and national and global contexts. Most of them rely on changing the reward structure so as to turn a situation into a Nash equilibrium. An obvious commonsense approach that we often adopt is to use social conventions, since these change the reward structure by adding the punishment of disapproval if they are not adhered to.

The disapproval does not have to come from others. Most of us are trained from childhood to feel bad about ourselves if we have done something that goes against that training, and this feeling can be strong enough to stop us from doing it. This constitutes a powerful force, and adherence to the social norms we have been taught is a major factor in a stable society. Even if others don't scold us, there is always that secret shame.

Unfortunately, it is not a shame that we can always rely on. I was brought up as a strict Methodist within a social group that strongly disapproved of drinking and dancing. Puberty saw to the latter, since the developing sexual urge was more than sufficient to overcome any shame that I might have felt about holding a girl close while dancing. University saw to the former, since the desire to be accepted by my beer-drinking peers meant that the reward for joining in their drinking was greater than any feeling of shame that I might have had.

Even so, social conventions can be very powerful. Witness the obedience of most of the male passengers to the women-and-children-first policy when it came to loading the lifeboats as the *Titanic* sank. Even then, one male passenger seems to have made it into a lifeboat dressed as a woman. That's the problem with social conventions: they may be powerful, but there is no *guarantee* that they will be adhered to. Pressure from society is not always as strong as pressure from rational self-interest.

This applies even when the social convention has been translated into law, such as the one that requires us to drive on the right-hand side of the road. This works well enough in the United States, where it puts us into a cooperative Nash equilibrium that provides us with safety, and anyone who takes it into their head to deviate from it risks serious injury or death. The situation can be very different in other countries, though, as I discovered when I was a passenger in a car in India and looked up to see a truck loaded high with vegetables swaying wildly as it bore down on us while travelling along the wrong side of the dual carriageway. The truck driver was trying to save time by cutting across a break in the central reservation instead of travelling past his destination to a point where he could make a legal U-turn and come back along the correct side of the road. When

I climbed back off the floor and took my hands from my eyes, I found that my driver had chosen the best available Nash equilibrium, coordinating his strategy with that of the truck driver by steering our car up onto the pavement and staying there until the truck had passed.

The problem was that the truck driver's idea of rational self-interest was rather different from mine or my driver's. This points up one of the main problems in applying game theory to real life, which is the assumption that the other person's rationality is the same as your own. It is not an insuperable problem, but it can certainly lead to some tricky situations.

I was once confronted in a Sydney pub by an inebriated soldier waving a gun after I had accidentally knocked a glass of cold beer into his lap. His behaviour was hardly rational, but a modern game theorist might have been proud of my solution, which was to appeal to the rationality of his still-sober friends (and to reach a co-ordinated agreement) by screaming 'Hold his arm!' as I dived for cover under the nearest table. Fortunately, they did.

We may sometimes act irrationally, but rationality is still our starting point. It is, after all, the feature that is supposed to distinguish us from other species, and it usually helps us to reach coordinated agreements if we are able and willing to communicate. Social conventions and social clues can help to maintain those agreements, especially if they are reinforced by a feeling on both sides that the agreement reached has been a fair one. As I show in the next chapter, though, just reaching fair agreements can be a tricky business, even when we use a strategy that is as simple and obvious as I Cut and You Choose.

2

I Cut and You Choose

ONE OF OUR MOST POWERFUL CHILDHOOD DRIVES is a sense of fairness, which we carry through into adulthood as a sense of justice. These senses were my first point of call in my search for tools that could help to promote and maintain cooperation. If an agreement to cooperate seems fair to all sides, I thought, then surely the parties will be less inclined to break it.

The sense of fairness seems to be deeply ingrained in our psyche and may come from a long way back in our evolutionary history. Monkeys have a sense of fairness, for example. Brown capuchin monkeys get frustrated and angry when they see others receiving better rewards for performing the same task. Researchers have found that they will sulk, refuse to do the task any more, and even throw their food rewards at the researcher in frustration, just as I once threw a bowl of fruit and custard at my mother because I thought my brother had got more than his fair share of this, my favourite dessert.

What could she have done to ensure that I was not envious of my brother's portion? The obvious answer would have been to use the I Cut and You Choose strategy, in which one of us divided the pudding into two portions and the other then chose which portion to take for himself. It may not have worked too well for us in practice, since I was only four at the time and my brother was only two, but game theorists have shown that this sort of procedure is the most equitable way in principle to distribute *any* finite resource so that the result is *envy free*. This is because the cutter has every incentive to divide the resource as equitably as possible while the chooser can't complain because he or she was the one who made the choice.

One of my first experiences of this strategy came on the day when I launched a rocket into my grandmother's bedroom. It was a big blue rocket, and it cost three times as much as the red firecrackers that accompanied it into the flames when I accidentally kicked my brother's box of fireworks into the family's bonfire during a holiday celebration. The fireworks went off with a splendid explosion that would certainly have woken Nanna, sleeping peacefully in her room. The rocket got there first though, carving a golden path through the air, passing through her open door, and lodging itself under her dressing table. It fizzed and spluttered briefly before exploding in a shower of blue and white sparks that brought her out of bed with a speed that belied her seventy-odd years. She appeared at the door brandishing her stick and mouthing words that I never thought she knew. It wasn't the stick that hurt, though. It was my father's declaration that I had to give half of my own box of fireworks to my brother.

I was only seven at the time, but even though I did not have the benefit of my later study of philosophy I still came up with what seemed to me to be some pretty good arguments. I pleaded that it wasn't fair, that tripping over his box wasn't my fault, that he shouldn't have put it so close to the fire. My father was adamant. The only concession I could wring out of him was that I should divide my fireworks into two piles and then my brother could choose which pile to take.

I made my selection with great care, determined that whichever pile my brother chose, I should not end up as the loser. It was the best that I could do. It was also the best that he could do. If either of us had insisted on more, my father had threatened to give all of the fireworks to the other one. Although I did not know it, my commonsense strategy of I Cut and You Choose was just the one that game theorists would have recommended in response to my father's strategy. (I discuss other strategies that my father could have adopted in chapter 5.) It was a simple application of the principle that they had christened *Minimax*.

Minimax means looking at a situation to see how much you might lose and then planning your actions so as to minimize that loss (that is, *mini*mising your *max*imum possible loss). It is the principle that Adam and Eve would have been well advised to adopt in the Garden of Eden, instead of risking losing the whole garden by indulging their curiosity about the taste of apples. We are also attempting to minimise our maximum possible loss when we take out insurance on a house or a car, reasoning that it is better to accept the cost of the premium rather than risk a larger, maybe catastrophic loss if we are involved in a car crash or if the house burns down.

I Cut and You Choose is a Minimax procedure because the cutter has every incentive to divide the resource equitably so as to lose as little as possible (the Minimax principle at work), while the chooser will obviously choose the piece that fits the same principle in her eyes. Its appealing fairness makes it an obvious candidate as a strategy for the cooperative sharing of resources in this troubled world. One common example is the division of property in divorce cases, which is usually done at present by assigning cash values to assets and then dividing the total cash value in some proportion. Game theorists have demonstrated that I Cut and You Choose would allow for other values, such as emotional attachments to particular objects, to enter the equation in an equitable way, which would be to everyone's advantage.

I Cut and You Choose has even been incorporated into some international treaties. The 1994 United Nations Convention on the Law of the Sea, for example, incorporates it into a scheme that is designed to protect the interests of developing countries when a highly industrialised nation wants to mine a portion of the seabed underlying international waters. The country seeking to mine would divide that area into two portions. An independent

---

> BOX 2.1

MINIMAX

Minimax is a new name for an old idea. Its essence is reflected in the old proverb 'half a loaf is better than none'. The comic novelist and bridge expert S. J. Simon described it in his book *Why You Lose at Bridge* as the principle of aiming for 'the best result possible' rather than 'the best possible result'. With this description he got Minimax down to a T.

The power of the principle was discovered by John von Neumann during his pioneering studies of game theory – a theory that he developed because he wanted to win at poker. In von Neumann's terminology, poker is a *zero-sum* game, because the gains of some players must come from the losses of other players, so the total gains and losses at the end of the game

>

agency representing the developing countries would then choose one of the two tracts, reserving it for future use.

It sounds brilliant in theory, and a great poke in the eye for the selfishness of the developed nations. When I experimented with the strategy, though, I found that it faced three major difficulties. The first was that different people can have very different values, which is not a problem in itself, but which can make it very difficult to assess and compare these values. The second difficulty is practical implementation, especially when more than two people are involved. The third and most serious difficulty is how to get people to accept the outcome when there is no independent authority to stop them from trying to get more than their fair share by cheating or bullying.

add up to zero. The phrase has gained some popularity with headline writers, but zero-sum interactions are not in fact very common in real life. In the early days of game theory, though, they were the only situations that it could handle. Von Neumann and his co-author, the economist Oskar Morgenstern, analysed the best strategies for winning at such games in one of the most unreadable books in history: *Theory of Games and Economic Behavior*, a 648-page tome, heavily laden with mathematics, which in my own library now serves as a doorstop, having been replaced on my shelves by later, more accessible works.

Their conclusion was that the Minimax principle *always* leads to the best strategy – for both sides! Unfortunately, this conclusion applies only to zero-sum situations, where the gains and losses balance out neatly. Such situations are rare in real life. When a thief smashes your car window to steal your radio, for example, he may make some money selling it later, but your

➤

Different Strokes for Different Folks

My first experiment with people's values wasn't meant to be an experiment at all, and the surprising result was pure serendipity. I was at a party where a plate of cake slices was being passed among the guests. When there were just two slices left, I took the plate and politely offered it to a fellow guest, who promptly took the *smaller* of the two pieces that were left, leaving me with the larger. That wasn't what game theory had led me to expect at all, since it assumes that people will always respond in the way that benefits themselves most.

Sometimes that response will be pre-emptive, in response to an expected action. On this occasion it was direct: I offered the two pieces of cake, she responded by taking the smaller one.

How could this possibly have benefited her more than taking the larger piece? There was only one way to find out, and that

loss (or that of the insurance company) can run to hundreds or even thousands of pounds. No balancing of gains and losses there. When business competition bankrupts one competitor while marginally increasing the profits of another, the balance is surely negative. It is negative for all parties in situations of conflict, from divorce to civil war.

Minimax can still be useful in such situations – it's not a bad idea to use business strategies that minimise your chance of bankruptcy, for example – but it can't be *guaranteed* to produce the optimum result. You might do better by gambling on a large return if the risks are low. When it comes to games with agreed sets of rules, though (such as poker and baseball), Minimax *is guaranteed* to give you the best chance. But how should you go about achieving the best result possible rather than chasing after the best possible result?

was to ask her why she had taken the smaller piece. Her answer was very revealing. She said that she would have felt bad if she had taken the larger piece. The benefit she would have gotten from taking the larger piece (in terms of satisfying her own hunger or greed) would have been more than offset by the bad feeling she would have had about herself for being seen to be so greedy.

So the assumption of game theory was right in this instance, once all of the factors had been taken into account. My fellow guest *had* taken the action that was of the most overall benefit to herself. Game theorists call that sort of overall benefit *utility*.

If they could measure it accurately, in the way that physicists measure the speed of light or chemists measure the concentrations of solutions, they could compare the values of rewards for different strategies, and game theory might become an exact science. As things are, game theorists have to resort to measuring

The answer, proved by von Neumann to be the optimum one, is often to use a mixed strategy, which means mixing up your actions or responses so as to minimise your possible maximum loss by not being too predictable. Baseball pitchers do this instinctively when they use a combination of fastballs, sliders, and curveballs during an important inning. But do they always get the proportions right? There are many possible permutations, but von Neumann proved that there is always just one that is optimum. This need not be mixing the pitches randomly and in equal proportions, because some will be more rewarding than others. A particular pitcher's fastball may be stronger than others, for example, and less likely to be hit. If he throws it all the time, however, it will become predictable and more likely to be hit, so it pays to mix in some of the weaker pitches. Von Neumann's mathematics allows us to predict the correct mixture,

➤

devices that help them to make comparisons but that might not tell the whole story.

One of those devices is to assign a cash value to the benefit. This may not be as difficult as it sounds. Our local corner shop, for example, charges around 5 per cent more for most goods than does the big supermarket a couple of miles down the road. They are still in business after many years because the locals find it more convenient to shop there, at least for small items. We can assign a cash value to that convenience in terms of the higher prices that they are willing to pay.

We assign monetary values to otherwise intangible benefits in many areas of life – in fact, this is what the modern science of economics is largely about. I have to admit to doing this when my children were young and I was trying to persuade them to clean up their rooms. Moral arguments weren't very effective, and neither was leading by example. What really worked was a

---

but I have been unable to discover whether any baseball teams are now taking advantage of it.

In sports where the mathematics has been compared with intuition, it has been found that intuition produces results that are in close accord with the Minimax principle. Take football. Economist Ignacio Palacio-Huerta from Brown University, a football enthusiast, watched over a thousand penalty kicks taken in professional matches in England, Spain, and Italy and analysed them in terms of a two-person, zero-sum game. Both the penalty-taker and the goalkeeper have to decide which way to shoot or dive respectively, and each will be stronger on one side than he is on the other. If neither has a clue as to what the other is going to do, each should choose to play his strong side. But neither player can choose his strong side all the time, because then the other player will quickly figure out from previous matches that

➤

bribe. The cost to me was negligible in terms of my income, but the gain to them as a percentage of their income was considerable. What I was really paying them for was giving up their play time for a while, and the amount that they were willing to accept reflected the value that they attached to that play time.

The same principles apply to some of the wider problems that we are faced with. In England, for example, tourists value the beauty of a countryside where the fields are divided by hedges. Farmers, however, have been busy digging up hedges to make larger fields. The answer? Find out how much the farmers would have to be paid to *stop* digging up their hedges and then pay them with money derived from tourism.

On a larger scale still, we are faced with the prospect of worldwide ecological catastrophe if habitat destruction goes on at its present rate in places like Brazil and Indonesia. But how much would *you* be willing to pay (as extra taxes, say, to support

this is what his opponent might be expected to do, and will react accordingly. In game-theory terms, each player must mix his strategies up to maximise his expected pay off (for the shooter, the probability of scoring; for the keeper, the probability of preventing a goal). According to the Minimax principle, the players should mix them up so that their expected pay off (success rate) will be the same whether they aim or move to the right or to the left, doing so randomly from game to game but in the appropriate proportions according to their strengths. When Palacio-Huerta analysed his observations, he found that almost all goalkeepers and shooters were superb exponents of game theory, choosing to aim right or left with appropriate frequencies.

overseas aid) to stop a Brazilian farmer or a logging company from clearing rainforest for farmland? How much would you be willing to pay to stop the clearing of rainforests in Indonesia (habitat of the endangered orangutan), which is currently happening over a large area so that palm trees can be planted to provide cheap palm oil for Western markets? How much would the producers have to be paid to stop these activities? Do these two figures coincide, or are they wildly different?

By looking at problems in this way and attaching cash values to otherwise intangible things, such as natural diversity, we can at least get a handle on the scale of the problem and what might need to be done to solve it. One of the difficulties, though, is that the size of the handle might change. When I paid my children to clean their rooms, for example, it worked well for a while, but then they began to *expect* bribes, and things escalated, just as they have in parts of the world in which bribery of officials is an accepted way of life. This was when I learned the practical difference between the sort of strategies that work well as a one-off and the sort of strategies that work best with repeated interactions – but more on that in chapter 5.

Bribes might sound like bad news, but game theorists have shown that they are an essential component of cooperation, although they usually prefer to call them by less pejorative names such as *inducements*, *rewards*, or *side-payments* (this latter is the correct technical term). Whatever you call them, they are payments (in terms of money, material goods, or even emotional support) that some members of a group have to offer to others in order to ensure a binding commitment of that person to the group. It sounds like a cold-hearted way of looking at things, but it can provide a clear-sighted view of what is going on behind the scenes in even the most emotional of circumstances. When my first marriage broke down, for example, a counsellor sat us down together and asked each of us whether the other person was offering enough to keep them in the marriage. After getting an answer, she turned to the other person and asked whether they were willing to offer more to save the marriage.

She wasn't talking about money but about respect, emotional support, and the whole host of things that make up a successful marriage. In doing so she was also implicitly treating human interactions as games that we play, with strategies and outcomes, gains and losses, winners and losers. This is nothing new to psychologists, and it does not necessarily devalue relationships – it merely looks at them in a different and often illuminating way. Game theorists use a similar model of human behaviour to compare the outcomes of the different strategies we use as we play the game of life, and to find out which strategies are best for different situations. At the very least they aim to list the outcomes of those strategies in rank order (bad, good, better, best, for example). To get full worth from their methods, though, they need to be able to attach numerical values to these outcomes.

Sometimes this can be done by attaching realistic cash values, but often it cannot. To overcome this problem as best they can, they have coined one of the ugliest words in the English language – *util*. A util is simply a number that expresses the relative utility of an outcome when that utility can't be interpreted in terms of money. It seems like a pointless exercise, but it actually permits comparison of the outcomes of strategies when money is not an appropriate or usable measure.

When we are asked to score preferences on a scale of 1 to 10, we are actually scoring in utils. The result of my cake-sharing experiment made sense as soon as I asked my fellow guest to score her preferences in this way. I first asked her to score the two pieces of cake as though she had been buying them in a shop where they were both the same price and where they were neither a bargain nor too expensive. She scored the larger piece at 5 and the smaller piece at 4. I then asked her to score her feelings about taking the smaller or larger piece on the same scale. She scored the smaller piece at 8 (it was a very nice cake) and the larger piece at 4. Treating these scores as utils and adding them up, she scored a total of 12 for the small piece and 9 for the large piece. This clearly illustrated her preference for the small piece. Perfect!

I repeated my experiment at other parties with other trays of cakes, and with trays of drinks. The results were almost always the same, and they didn't depend on whether it was a man or woman that I was offering the tray to – both seemed to gain more utils from taking the smaller piece of cake. This was confirmed when I asked them to assign numerical values. What may have made a difference was the country I was in, which happened to be England, where this sort of politeness is highly regarded. I repeated the experiment in Australia, though, and obtained the same result – except when I offered the tray to my brother, who promptly took the largest piece of cake with a big grin on his face. How I felt about it didn't bother *him* – the size of the cake took precedence. (Maybe he was also getting back at me for the fireworks.)

The Cake-Cutting Problem

When I looked into the matter further, I discovered that attaching numerical values to human feelings is just one of the problems that we face in working out how to divide a finite resource in a fair, envy-free manner. A second problem is finding a workable formula to produce such a division. This is known as the *cake-cutting problem*, and a full general solution evaded mathematicians until well into the twentieth century.

An ancient group of rabbis found a solution to a particular case without the aid of modern mathematics, though, when they were confronted with the case of a man who had three wives. Their solution appears in the Babylonian Talmud.

The wives weren't actually the problem; it was how the man's estate should be shared among them when he died. Each of them had legally binding nuptial agreements (unlike some celebrities today), but the three agreements were different. One wife's agreement specified that she should receive 100 dinars from his estate (approximately £5,500). The second wife, who seems to have had a better lawyer, was to receive 200 dinars. The third wife, who had the best lawyer of all, was owed 300 dinars.

The rabbis had the job of coming up with a *mishna* (a brief set of conclusions) that would provide guidelines on what to do if his estate amounted to less than the required 600 dinars. How could it be divided up in the fairest way that still accorded with the spirit of the different marriage contracts? After due consideration they came up with three different recommendations, depending on what his estate was worth. Two of the recommendations made intuitive sense, but the third puzzled Talmudic scholars until very recently.

If the estate was worth 300 dinars, they recommended proportional division (50, 100, 150), which satisfies the ratios specified in the marriage contracts. If the estate was worth only 100 dinars, the sages decided that equal division would be a fairer split. What scholars could not understand until 1985 was why the rabbis had recommended a 50, 75, 75 split if the man left an estate worth the intermediate amount of 200 dinars. The recommendation did not seem to make any sort of sense, and many scholars dismissed it outright. One even claimed that since he could not understand it, it must be a mistranslation. Then the problem came to the attention of the Nobel Prize-winning game theorist Robert Aumann, who in collaboration with economist Michael Maschler used game theory to prove that the rabbis involved in the original discussion had brilliantly hit upon the optimum, fairest solution to the problem.

The argument that they presented is both beautiful and simple. They began by considering the problem of how to divide a resource when one person claims ownership of all of it and another claims ownership of half of it. The answer? Divide it according to the ratio 75:25, because the ownership of half the resource by one of the parties is undisputed (and goes to that party), leaving the other half in dispute, for which the fairest solution is to divide the second half 50:50. They called their solution 'equal division of the contested sum', and proved that in the case of the man with three wives 'the division of the estate among the three creditors is such that any two of them divide the sum they together receive, according to the principle of equal division of the contested sum'.

It sounded to me as though this would be an excellent principle to apply to sharing in everyday life, first because it is so simple and second because it feels so fair. I had the opportunity to try it out when a friend and I went to a garage sale and found a stall loaded high with second-hand books. Rather than competing for the most desirable books, we pooled our resources and bought all of the books that either of us wanted. Then we divided them into three piles: the ones that I particularly liked but he didn't want, the ones that he particularly liked but I didn't want, and the ones that we both wanted. We then took turns in choosing a book at a time from the third pile (the contested sum) until we had divided it equally. Very simple. Very satisfying.

Equal Division of the Contested Sum can even be applied to global problems. It is now being looked at seriously, for example, as the fairest way of settling territorial disputes. It might be worth applying it to the present dispute concerning oil exploration rights on the Arctic Lomonosov Ridge: just grant each country the rights to the uncontested bit of its claim and divide the rest equally among them (see the note on the 1994 UN Convention on the Law of the Sea). Admittedly such problems are very complex, but as a scientist I can certainly see the appeal in such a simple solution. It might work.

The I Cut and You Choose strategy is a scaled-down version of Equal Division of the Contested Sum, since it produces equal division when there is no uncontested sum to worry about so everything is up for grabs. This is by no means the end of the matter, however, as my brother and I discovered when my father used this strategy to divide up the household jobs. He would write out a list of them (putting out the rubbish, washing up, sweeping the floor) and get one of us to divide it into two lists that we thought were equal. The other could then choose which list to have. Just to make sure that there were no complaints, he alternated the lister and chooser each week.

So far, so fair. But we also had a younger brother, and when he became old enough to do his share of the jobs, all hell broke loose. With three of us to share the jobs, it seemed impossible to divide up the list of chores and share out the three sub-lists without arguments. We never managed to balance the lists fairly without subdividing some jobs, and even then there were arguments that led to further subdivisions, and so on in a process that seemed to have no end.

We didn't know it, but we were replicating some of the early efforts (and dilemmas) of mathematicians to solve the cake-cutting problem when more than two people are involved. One of the problems (even with a cake) is that the initial division into three parts is bound to produce slightly unequal segments in practice. This means that the first person to make a choice can pick the larger one, to the envy of the other two.

The earliest attempts to resolve the problem produced a complicated procedure that started with the person who took the first choice (and the largest slice) being required to cut a sliver from it to be further subdivided. Unfortunately this procedure produced an infinite cascade of division and subdivision, as happened with my brothers and me. It wasn't until 1995 that Steven Brams from New York University and Alan Taylor from Union College came up with a practical solution that had a finite number of steps. Their calculations were cumbersome, but they were manageable with the aid of computers. Brams and Taylor subsequently patented a procedure for the fair allocation of multiple goods based on the concept of the 'adjusted winner'. Their basic principle was to take account of the fact that different people can attach different values to the same assets, so a division between two parties, for example, can be worked out in such a way that each party gets more than 50 per cent *as they perceive it* – a win-win solution if ever there was one, equally applicable to birthday parties and legal parties! Their method (now licensed to *Fair Outcomes Inc.* at www.fairoutcomes.com) and its potential applications are described in their book *The Win-Win Solution: Guaranteeing Fair Shares for Everyone*.

One of those applications is in the negotiation of land rights and other territorial deals, in which considerable progress is now being made in working out more fair and equitable approaches. A more surprising application is to voting. In this case the problem of ensuring fair and equal representation in a democracy is simply the cake-cutting problem applied to millions of voters so that all of their votes have equal weight. Interestingly, no current voting system comes close to being representative when judged in the light of the Brams-Taylor solution. To give one example, the weight of individual votes in closely contested electorates is far higher than that of votes for the losing side when the election is more one-sided, which count for virtually nothing since that candidate is *never* going to win.

The Brams-Taylor solution provides a benchmark for fair division, though. The best that we can hope for in real life is that our solutions should come as close to this benchmark as practicable. My father eventually achieved this in dividing up the household chores by allowing each of us to make just one change to the three sub-lists, after which he shuffled the lists and gave us one each at random.

His solution worked because we didn't have much of a preference for one chore over another (we just hated them all). Random distribution isn't always the answer, however, as I discovered when I experimented with the best way to divide up a wedding cake. I carried out my experiment at a friend's wedding reception. The wedding was over, the speeches were over, and the cake had been cut into slices that were now being distributed. It was a beautiful chocolate cake, covered with layers of icing, and I was curious to see whether people would take the larger slices first. It seemed, though, that most people were more interested in the composition of the slices than they were in their size. Some were choosing slices that had the most icing, while others were eyeing the icing with distaste and going for slices that had more cake. The 'fair' division of the cake into approximately equal slices, with each slice consisting of a piece of cake covered with icing, had not completely satisfied any of them.

People at other tables were leaving chunks of icing or pieces of cake on the sides of their plates (when the reception was over I counted thirty-one chunks of leftover icing and seventeen pieces of leftover cake). Some people at my table had started swapping their pieces of icing for a piece of their neighbour's cake when I suggested that we turn it into a community effort. I got them to separate the icing from the cake and put the separate pieces on a large plate. We then passed the plate around the table with everyone choosing one piece of cake or one chunk of icing at a time until there was no cake or icing left. It was that simple. No one was unhappy with the outcome, and more than half the people at the table said that they had done better than their original choice.

My experiment suggested that subdivision by the people who are going to do the choosing is the best practical approach to sharing out a resource when different people have different preferences for different bits of it. I was interested to find out, from a friend who works in foreign aid, that this is just the way that some villagers share out aid among themselves. One person may end up with blankets, and another with food, for example, when the aid is initially distributed under often-chaotic conditions. They could swap with each other but find it much more effective to keep what they really need and lump the rest together, with each choosing successively from the pile. The potlatch ceremony of the indigenous peoples of the Pacific Northwest performs a similar function in redistributing wealth, with the interesting variation that prestige can be counted as one of the goods, since those who contribute most to the communal pile gain the most prestige.

Democratic sharing isn't always so easy to achieve, as I found when I was the policy coordinator for a newly formed, and now extinct, political party in Australia. One of the reasons for our extinction was our keenness to be truly democratic. Every policy decision had to be discussed, decided, and agreed upon democratically by the whole of the membership. This took an unconscionable amount of time and a huge amount of administration, and often resulted in watered-down or even self-contradictory policies.

I decided to try an experiment in making things easier for the members (and for myself) by introducing a decision-making method called the *Delphi technique*. The idea has game theory roots, and it is very simple in principle. Everyone has their say (about policies in this particular case) in a questionnaire, and then an independent facilitator (me again, in this case – we were a very small party!) summarises their arguments and conclusions and sends the summary back out to all members of the group. Everyone can then vote again after they have considered and revised their arguments and conclusions in the light of what the others have said.

The idea is for the members of the group to use the best information available to them to converge on the best decision. Businesses use it for market forecasting, since it can be reasonably argued that the averaged opinion of a mass of equally expert or equally ignorant observers is more reliable as a predictor than the opinion of a single randomly chosen observer. Author James Surowiecki provides an entertaining example in *The Wisdom of Crowds* when he points out that the TV show 'Who Wants to Be a Millionaire?' pitted group intelligence against individual intelligence, and that every week, group intelligence won'.

When I tried to use it to make political decision making as democratic as possible, however, the members didn't like it at all – not because it was unfair but because I had introduced it without consulting them! But how then was I to consult them about how I should consult them? Caught on a sinking ship in their whirlpool of logic, I followed the only course possible: I jumped ship and left them to it. I have had no direct involvement with politics since.

My brief excursion into politics, however, was connected to my deep concerns about the direction the world was heading in. I now know that politics isn't my forte. For one thing, I retain my childhood sense of fairness and fair play, which is not something that fits very well with practical politics. But I have never stopped thinking about the issues that I entered politics to address, in particular how to promote and maintain cooperation, justice, and fairness.

My investigation of I Cut and You Choose revealed that it can be a very effective strategy for fair sharing but that it usually requires enforcement by an external authority to make it work (as in my father's division of the fireworks). Fairness itself does not provide guaranteed self-enforcement of cooperative agreements when it comes to the practical politics of everyday living. I needed to look further for strategies that would carry their own enforcement. Before doing so, I decided that it was time to look more closely at the logic that draws us into social dilemmas and to see if I could garner any clues for new cooperative strategies from the nature of the logic itself. When I did, I discovered that there is not just one social dilemma, but seven, waiting to snare us in our everyday lives!

3

The Seven Deadly Dilemmas

THE PRISONER'S DILEMMA IS just one of many social dilemmas we come up against in our attempts to cooperate. Seven of these are particularly damaging, and game theorists have given each an evocative name. One is the Prisoner's Dilemma. The other six are:

- The *Tragedy of the Commons*, which is logically equivalent to a series of Prisoner's Dilemmas played out between different pairs of people in a group.

- The *Free Rider* problem (a variant of the Tragedy of the Commons), which arises when people take advantage of a community resource without contributing to it.

- *Chicken* (also known as Brinkmanship), in which each side tries to push the other as close to the edge as they can, with each hoping that the other will back down first. It can arise in situations ranging from someone trying to push into a line of traffic to confrontations between nations that could lead to war, and that sometimes do.

- The *Volunteer's Dilemma*, in which someone must make a sacrifice on behalf of the group, but if no one does, then everyone loses out. Each person hopes that someone else will be the one to make the sacrifice, which could be as trivial as making the effort to put the rubbish out or as dramatic as one person sacrificing his or her life to save others.

- The *Battle of the Sexes*, in which two people have different preferences, such as a husband who wants to go to a football game while his wife would prefer to go to a movie. The catch is that each would rather share the other's company than pursue their own preference alone.

- *Stag Hunt*, in which cooperation between members of a group gives them a good chance of success in a risky, high-return venture, but an individual can win a guaranteed though lower reward by breaking the cooperation and going it alone.

In a sense, all of these dilemmas are the same dilemma. Cooperation would produce the best overall outcome, but the cooperative solution is not a Nash equilibrium and there is at least one non-cooperative Nash equilibrium just waiting to draw us into its net. Here I investigate how the traps work and how they affect us in real life. In the following chapters (which can be read independently of this one) I look at ways in which we might avoid them or escape from them.*

---

*For simplicity, I concentrate on situations in which each party has to decide on its strategy without knowing what the other party has decided to do. Game theorists call these *simultaneous games* (as opposed to *sequential games*) and represent them as matrices like that shown in chapter 1 for the Prisoner's Dilemma, but

My first point of call was one of the most widespread and perplexing problems that game theory has uncovered.

The Tragedy of the Commons

The Tragedy of the Commons (Box 3.1) is a social dilemma on a grand scale. When I first had the idea of writing this book, I began to collect examples from newspaper stories, and my study floor rapidly became covered with pile after pile of newspaper clippings. The stories cover DVD piracy, benefit cheating, copper stolen from Russian power plants, overfishing, e-mail spam, and side benefits from credit cards at the expense of those who are caught in the credit trap. They also cover resource depletion, pollution, and global warming. All are examples of the Tragedy of the Commons, which is a multiperson Prisoner's Dilemma in which the cumulative effect of many people cheating on cooperation can eventually be catastrophic.

My wife Wendy and I saw a real-life example when we visited Sri Lanka just after the devastating tsunami of 2004. Funds had been donated to help people who lived in the affected areas to move out or to rebuild their damaged houses. A local guide told us that some people from outside these areas were actually moving *into* them so that they could claim a share of the benefits. In doing so, each could be seen as taking a small slice of the funds from each of the people already there. Overall, it set the scene for a Tragedy of the Commons: if too many people had pursued the same grasping strategy, there would not have been enough

---

with different combinations of rewards, strategies, and outcomes. These matrices are a convenient shorthand for visualising what is going on, and they provide handy points of reference, but they are by no means essential and can be bypassed by readers who do not find them helpful.

> BOX 3.1

THE PRISONER'S DILEMMA AND
THE TRAGEDY OF THE COMMONS

The Prisoner's Dilemma arises from the presence of just one Nash equilibrium. All of the other dilemmas featured in this chapter involve at least two Nash equilibria. To make the different dilemmas easier to understand and compare, I have portrayed them all in terms of positive rewards. In the Prisoner's Dilemma (as described in chapter 1), for example, the maximum sentence was ten years, and the reward for a particular strategy was the number of years that could be shaved off the maximum. If Bernard and Frank both confess and go to jail for four years, for example, their reward is essentially six years.

Drawing their matrix in this way, we get the following:

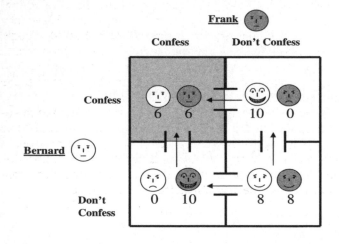

The square with the grey background represents a Nash equilibrium in that neither Frank nor Bernard can independently improve his situation without the other then undermining his attempt. The best all-round result would be the bottom right-hand cooperative situation, but through each trying to improve his own position they get stuck in the top left-hand Nash equilibrium.

The Tragedy of the Commons is essentially a multiperson Prisoner's Dilemma. Our choice of strategies is to cooperate with the group by taking no more than our fair share, or to cheat by using more than our fair share of a communal resource. The results can be very different, depending on what the other members of the group choose to do.

Australian vegetable farmers, for example, are currently restricted in how much water they are allowed to use as a result of a serious drought in the country. If they cooperate with the restriction, they will get a lower yield per acre – for the purposes of illustration, let's say 5 tons per acre rather than the normal 10 tons per acre. If just a few cheat by using water freely, they could still get 10 tons per acre. If most of them cheated, though, the reservoirs would run low and their yields would drop, say, to 2 tons per acre. More severe restrictions would also come into force, and individuals who cooperated with the new restrictions might only then get 1 ton per acre.

The outcome depends on how most of the farmers see themselves. If they think of themselves as members of a cooperative group, they are likely to cooperate. If they see themselves as competing individuals, each trying to do the best for themselves even if it is at the expense of others, then their game theory matrix will look like the diagram on the following page.

➤

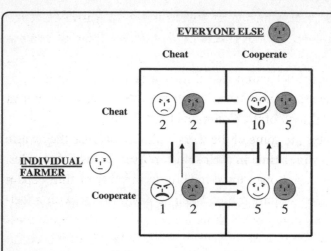

In other words, once they see themselves as individuals, their dominant strategy is to cheat, regardless of what the strategy of the others is. When they all cheat, though, then (just as in the Prisoner's Dilemma) they all end up in the top left-hand corner rather than the bottom right-hand cooperative corner. The key to cooperation is to find some incentive that rewards people (either psychologically or in practical terms) for being loyal and cooperative members of the group. This is a key that I explore further in later chapters.

money for *anyone* to build a proper house or rebuild a damaged one.

The Internet provides a less obvious example, but we experience the ghostly hand of the Tragedy of the Commons every time we use our computers to surf the Web. When we use it to download massive music, video, or game files, any one of our individual downloads has little effect. Together, however, they delay our e-mails, disrupt our Skype calls, and contribute to our premature strokes and heart attacks as we sit frustrated at the keyboard enduring the effects of yet another spike of congestion, otherwise known as an 'Internet storm'. We may not see our individual behaviour as selfish, but each of us is in fact trying to take a little more than his or her fair share, which is just what the Tragedy of the Commons is all about.

Spammers are some of the worst culprits, because they waste so many people's time in their selfish pursuit of just a few sales. Every morning, I open my e-mail to find twenty or thirty spam messages clogging my inbox. I delete most of these with a feeling of irritation, but one did make me laugh. It was a spam message, presumably sent to millions of people, offering a cheap anti-spam filter for sale!

Internet storms and spam messages are trivial problems when compared to resource depletion, global warming, terrorism, and war, but all have their origins in the same frustrating game of logical table tennis – the oscillating choice between cooperating with others or going our own way, and to hell with the others.

The Free Rider

The Free Rider dilemma, like its close cousin, the Tragedy of the Commons, is a multiperson Prisoner's Dilemma. Some common examples are: leaving a mess for others to clean up in shared accommodation; the choice between remaining seated and standing to get a better view (while blocking the view of others) at a sporting event or an outdoor concert; refusing to join a trade union but still accepting the benefits won by its negotiations with employers; credit-card fraud (because the losses to the supplier are passed on to honest consumers as higher prices; burglary; and even disarmament (if a majority of people wanted their country to disarm, they would still benefit from the military protection provided if a minority wanted to keep the country armed and were willing to commit resources to it).

The choice between cooperating with others and pursuing our own interests regardless of others is one that we regularly face when it comes to the care and use of communal resources. It can often seem to us that a free ride doesn't really cost anyone anything. For example, a friend of ours rented a skip to get rid of some rubbish and was highly indignant when some of her neighbours dropped small items of their own into it. 'But what's the problem?' they asked. 'You had to rent it anyway, so our little bit of rubbish wasn't costing you any extra.'

Their logic was hard to dispute – in fact, it was impossible, because it was the same logic that underlies the Prisoner's Dilemma. This is not surprising, given that the Free Rider problem has a similar logical structure to the Tragedy of the Commons (Box 3.2). It is similarly hard to resolve, because the free rider's strategy of making free use of a resource that would be there whether they used it or not seems to make perfect sense. So it does – until others follow their precedent. If the whole street had started to put unwanted material in my friend's skip, for example, there would have been no room left for her own rubbish, leaving her to wonder why she had bothered to rent it in the first place. In fact, if she had foreseen this behaviour, she wouldn't have rented it!

The Free Rider problem might more accurately be called the Easy Rider problem, because the cost to society is insignificant, but it is not zero. If too many people become free riders, the many insignificant little costs add up to a large burden. This was surely the case in the old Soviet Union when the citizens of

➤ BOX 3.2

FREE RIDER

Free Rider introduces a new and unexpected twist to the Tragedy of the Commons. Let's say a new steeple on a church would cost £100,000, and everyone is asked to contribute £100. I can work out the benefit to myself in dollar terms by asking myself how much I would be willing to contribute if it would make the difference between having a new steeple and not. Let's say the answer is £200. Under what circumstances should I then contribute, rather than letting others contribute and simply enjoying the benefits? From a self-centred point of view, a simple matrix of (benefit to me minus cost to me) helps me to work it out:

		EVERYONE ELSE		
		More than 1,000 people contribute	**Exactly 999 others contribute**	**Fewer than 999 others contribute**
ME	**Contribute**	100	100	−100
	Don't Contribute	200	0	0

Very interesting! There is only one situation where it is worth my while to contribute, and that is the point at which my contribution would make all the difference. Game theorists call it a point of 'minimally effective cooperation', and finding such points can be one of the keys to cooperation. Unfortunately, it is also the point beyond which it becomes worthwhile to be a free rider, relying on the contributions of others – which is fine until I realise that there are thousands of identical matrices, with someone else in the role of 'me', and I am lumped in with 'everyone else'!

It can be very difficult to identify the point of minimally effective cooperation in practice, which means that the pay off matrix usually looks more like this:

EVERYONE ELSE

		Enough other people act	Not enough other people act
ME	Act	Benefit-Cost of action	Cost of action
	Don't Act	Benefit without Cost	No benefit

It is clear from this matrix that the Don't Act strategy (that is, cheat) is *dominant* for each individual acting for themselves. Only if individuals see themselves as members of a group will the outcome be different. Just as in the wider version of the Tragedy of the Commons, the key to cooperation is to find some incentive that rewards people who see themselves in this way, either psychologically or in practical terms.

Moscow took advantage of free steam heating and controlled the temperature of their houses by leaving the heating on full and simply opening the doors and windows.

I discovered an interesting variant on this behaviour during a recent visit to Hungary, where many people are living in thin-walled apartments built during the communist era. They now own these apartments and have to pay for their own lighting and heating. The residents in the inner apartments, however, get a free ride during the winter, since the walls are so thin that the heat from the outer apartments permeates quite satisfactorily to keep them warm! Their free ride consists in the fact that those in the outer apartments are unintentionally paying for their heating.

Political scientists know the Free Rider problem as the Malibu Surfer problem, since many of the beach bums who ride the waves at Malibu are seen as free riders, living off social welfare. It can be argued that so-called Malibu surfers actually consume very little and that they use fewer resources compared to rich people, who also tend to maintain lifestyles that are less ecologically sustainable. This argument faces the counter-argument that the cost to the community of a few beach bums surfing the breakers in Malibu may be negligible, but if thousands of young people started to do this, the costs to the rest of us would soon mount up. Society can tolerate a few such free riders, but not a large number, however much we might wish vicariously to share their freedoms.

Young itinerants looking for excitement are not the only free riders. Historian Edward Gibbon referred to his tutors at Magdalen College, Oxford, as 'decent, easy men, who supinely enjoyed the gifts of the founder', which is surely an excellent definition of a free rider. In Australia we call them 'bludgers', a term that, in the working-class Australia of my youth, extended to anyone who did not perform a labouring job with their hands. Movement to a desk job was regarded as a soft option and treated with disdain, reflected in Australian poet Dorothy Hewett's immortal line 'The working class can kiss me arse I've found a bludger's job at last.'

Bludgers will probably always be with us. The main problem is to make sure that their numbers don't get out of control. But how can we do this? One approach is to make it progressively more risky, or more costly, for each additional free rider to enjoy the benefits of their ride.

I once shared a secretary who worked out a neat way to do this with members of our group who habitually left it until the last minute before giving work to her. They were effectively taking a free ride by putting pressure on her instead of taking the trouble to plan their time. She responded by putting a notice on her door that read: 'Lack of planning (on your part) does not justify an emergency (on my part).' Thereafter she might help one person with an urgent piece of work, but if a second person came along with a similar request, he or she would get a stern lecture, and probably a refusal. A third person, no matter how senior, would receive an outright refusal. Her strategy worked, and the number of urgent requests soon diminished.

Another technique for dealing with free riders is to change the reward structure so that the temptation to take a free ride is not there in the first place. A group of us who have formed a welcoming committee to provide new residents with social contacts in our village in Australia used this strategy with great success with a woman who gate-crashed our New Year's party.

We didn't know that anything was amiss at first. She simply came up to our table in a Chinese restaurant, apologised for being late, and sat down. We hadn't advertised the party, but we assumed that she was a new resident who had heard about it somehow. Only after she had gone, having drunk several glasses of offered champagne and ordered a special meal for herself, did we realise that she had also left without paying. The money that the rest of us had put into the kitty fortuitously covered her bill, but she left us in a pickle when it came to paying the waiter, who deserved his tip. So it didn't really cost us anything (although it cost the waiter something), and we wrote it off as a life lesson. We talked about what strategies we might use if such a situation arose again and got a chance to put our theory into practice when she turned up unabashed at one of our regular coffee mornings a couple of months later. One by one we managed to leave quietly as we finished our coffee, sticking her with the bill for all of us. She hasn't been back since.

Some of the examples I have given might seem rather trivial, but free riding is not always such a light affair. Sometimes it can have very serious consequences indeed. Global warming is one example: why not gain an economic advantage by letting other countries bear the cost of reducing carbon emissions? But when too many countries use the same logic, we all go under – metaphorically, and perhaps literally when the sea level rises.

Another example of the serious effects of the Free Rider problem in the modern world is that of corruption, which can even lead to the destabilisation of a country. The free rider is the individual official who takes a bribe or a kickback, leaving the other officials to maintain the law. But when too many of them start to think the same way, corruption balloons out of control, and the community services that the officials are supposed to be overseeing simply collapse. Maybe this is the sort of thing that led Peter Ustinov to comment that .corruption is nature's way of restoring our faith in democracy'.

Finally, I leave you with a poignant example from Chinese author Aiping Mu. In her book *Vermilion Gate* she tells the story of her childhood during the Cultural Revolution:

> During the "storm of communization," peasants put much less energy into working for the collective economy than for themselves, because the rewards were the same no matter how much or how little they worked, and no one could be bothered to take care of the collective property. The most painful experience was eating at the mass canteens, which were supposed to liberate women from daily

cooking and hence to increase their productivity and increase the quality of life. The outcome was just the reverse.

Misled by the propaganda, peasants assumed that a life of abundance had begun, and they could eat their fill. . . The peasants lost nearly everything, even their cooking utensils and food reserves. . . When the famine ended . . . one estimate put the number of deaths in rural China at 23 million.

Chicken

Sometimes we find ourselves in situations where the person who makes the first move loses out. The game theorists' evocative name for such situations is derived from the film *Rebel Without A Cause*, in which the characters Jim (James Dean) and Buzz (Corey Allen) play a game called 'chickie run'. They race stolen cars toward an abyss; the first one to jump out is the loser and called 'chicken'. The loser turns out to be Buzz, but in an ironic twist he loses even more when a strap on his leather jacket gets caught on the car door as he jumps, and he is dragged over the cliff with the car.

Unwillingness to lose out by being the first to make a move can sometimes have hilarious consequences. Naval commander Gaurav Aggarwal gives a wonderful example from his passing out parade at the Naval Academy, when he and the army general who was the guest of honour both stood frozen in mid-salute, each unwilling to complete the salute appropriate to his service until the other had backed down by completing the salute of his service. The position was only resolved when one of them started to crack a grin at the ludicrousness of the situation.

Laughter can take us a fair way in defusing some of the games of Chicken that we seem regularly to play. I have even used it to defuse a potential road rage situation, when another driver and I nearly collided as our two lines of traffic were merging on a country road in Australia, with each of us bent on getting in ahead of the other. Putting on my best English accent, I wound down the window and said, 'Please, after you,' with a smile. 'Stupid idiot,' he growled as he drove off – but at least he did drive off, while my Aussie passengers giggled in the back, nearly blowing my cover.

The problem in game theory terms (Box 3.3) is that there are two Nash equilibria, each favouring a different party – the party who doesn't back down. If the two pedestrians in my earlier story were walking towards each other along a pavement that was only wide enough for one, one of them would have to lose out by stepping into the gutter so that both could pass. The logical resolution, and Nash equilibrium, is still for one of them to step into the gutter, even though that person loses out by getting muddy boots. If both refuse to step into the gutter, though, the result will be an argument, and perhaps even a fight. In an equivalent situation between countries, it can even mean war.

In politics the situation is sometimes dignified by the term *brinkmanship*. Whatever the name, though, the parties involved find themselves with some unpleasant choices. If one gives way, the pair end up in a Nash equilibrium that strongly favours the other. If neither gives way, though, both find themselves in a situation that could be catastrophic. This came close to happening in the 1962 Cuban Missile Crisis, when the Soviet Union and the United States were on the brink of nuclear war after Khrushchev refused to remove Soviet missiles from Cuba and Kennedy refused to lift the U.S. naval blockade.

➤ BOX 3.3

CHICKEN

Returning to one-on-one situations, we come to the danger-ous game of Chicken. Here it is not as much a matter of assign-ing specific numerical values to rewards (which can be difficult in many cases) as of looking at how well you might do out of a situation in the order: good, neutral, bad, worst. Let's apply it to the two people walking along a pavement towards each other. For each of them the good outcome is for the other one to make way by stepping to one side; the neutral outcome is for them both to step to one side; the bad outcome is to be the one who steps aside; and the worst outcome is for both of them to refuse to step aside. The resultant matrix looks like this:

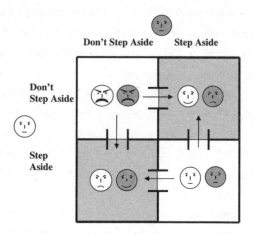

No need for numbers here – the expressions on the faces say it all. As the arrows on the diagram show, there are two possi-ble Nash equilibria. In each of them, one person is pleased with the outcome, but the other is unhappy. Either is preferable to

➤

neither person giving way (either on a pavement or in a more serious situation such as the Cuban Missile Crisis), but who is going to give way? It would be preferable if both did, but this needs coordination.

Note, too, that Hawk-Dove leads to a similar matrix of possible strategies and outcomes. It represents one of the most difficult and dangerous situations we can find ourselves in, and sometimes it seems as though no answer is possible. This can be true in one-off encounters, but as I show in chapter 7, there is a surprising solution if the parties know that the encounters are likely to be repeated in the future. In fact, repeated interactions turn out to be another key to resolving problems of cooperation.

◼

Bertrand Russell famously compared the behaviour of the two statesmen to a juvenile game of Chicken, with the future of the world at stake, in his book *Common Sense and Nuclear Warfare*:

Since the nuclear stalemate became apparent, the Governments of East and West have adopted the policy which Mr. Dulles [the secretary of state under Eisenhower] calls 'brinkmanship'. This is a policy adapted from a sport which, I am told, is practised by some youthful degenerates. This sport is called 'Chicken!'. . . As played by irresponsible boys, this game is considered decadent and immoral, though only the lives of the players are risked. But when the game is played by eminent statesmen, who risk not only their own lives but those of many hundreds of millions of human beings, it is thought on both sides that the statesmen on one side are displaying a high degree of wisdom and courage, and only the statesmen on the other side are reprehensible. This, of course, is absurd. Both are to blame for playing such an incredibly dangerous game.

We are not the only species to play the game. Many animals play it as well. Biologists know it as Hawk-Dove because, when it comes to competition for food, space, mates, or other indivisible resources, most animals tend to adopt either an aggressive Hawk strategy, or a Dove strategy by making a show of aggression but then running away.

In the natural world the two strategies correspond respectively to all-out aggression or ritualised displays of aggression. This is a very simplified picture, of course, but it reveals some essential truths, especially when it comes to working out which sort of strategy is best. The answer is neither! It turns out that the Evolutionarily Stable Strategy (the one that works best in the long run) is a mixture. For individual animals it means sometimes behaving aggressively and sometimes making a show of aggression. For populations it means some members adopting one sort of strategy and some another.

The proportions of the two types depend very much on risk (of injury during a fight) versus reward (for winning the fight). Male elephant seals are predominantly Hawks, and will risk serious injury during a fight because only the victor (the 'beachmaster') gets to mate with the females. Bullfrogs are also Hawks, but in this case only because they lack the capacity to inflict serious injury on each other. Scimitar-horned oryx, deer, and rattlesnakes, on the other hand, would run a high risk of death if they fought seriously and so have evolved a ritualised Dove strategy.

Most often, however, there will be a mixture of strategies within a population, as is the case with scorpion flies. The largest males are aggressive and will fight for dead katydids to offer to females, with high mating success as the result. Others are smaller and can do no more than produce saliva as a copulatory gift (the mind boggles at what the equivalent human behaviour might be!); their mating success is middling, but at least it is better than for the smallest flies, which can't produce enough saliva and have very low mating success. Game theory predicts that these three strategies will remain in balance within a population, and this is just what happens. If some of the flies in an upper echelon die off, some of those in the next lowest echelon will take the opportunity to change strategies until the strategic balance is restored.

When it comes to individuals, a mixed strategy is often the best theoretical option: sometimes putting on a show of aggression and sometimes actually carrying out the threat, as boxers and sumo wrestlers do. Game theory tells us that the mixture depends on the balance of danger and reward, and also that the threat must sometimes be acted on to make it credible. A boxer might feint and feint and feint, for example, but one of those feints is going to turn into a real punch. The opponent knows this and is forced to maintain his guard, just in case. If a boxer always feinted, the opponent would know this and wade right in.

Threats are useless, though, without credibility. I recently saw a woman in a supermarket screaming at a recalcitrant child, 'If you don't come here at once, I'll murder you!' The child, who will doubtless become a game theorist when she grows up, looked her mother in the eye and said, 'No, you won't,' before carrying on with her misbehaviour. The child realised that the threat was not credible. The mother should have realised this as well and substituted something more credible.

Threats may fail through lack of credibility because of a lack of communication. The film *Dr. Strangelove* turns on this point.

The Soviets believe that they have produced a credible threat in the form of a doomsday machine that will respond automatically without human intervention if the other side attacks. The irony is that the threat is not in fact credible, because the Soviets haven't had time to inform the United States about it before the mad Brigadier General Jack D. Ripper launches a nuclear attack on them.

On a lighter note (literally), in an experiment designed by Yale University game theorist Barry Nalebuff, a group of overweight people were photographed wearing tiny bathing suits and threatened that the photographs would be shown on TV and published on the Internet if they failed to lose fifteen pounds in two months. Now *that's* a credible threat!

The threat was particularly powerful because the dieters had voluntarily limited their own options. This is a very effective way to show that you mean business. One protestor against a road development through a site of special scientific interest in the United Kingdom, for example, handcuffed himself to the underside of a bulldozer so that his threat to sacrifice his life if the bulldozer were moved became a very real one indeed. He would have *had* to carry out his threat if the other party had dared him to (by starting up the bulldozer), which is the whole aim of limiting your own future options. Similarly, when Hern·n CortÈs landed with a fleet of eight hundred Spanish soldiers near the site of present-day Veracruz on April 21, 1519, he destroyed the ships in which he had arrived to send a message to his men that there would be no going back – forward was the only option. The onlooking Aztecs got the same message.

Limiting your own options, though, is not necessarily confined to such dramatic circumstances. As I write this I am sitting in a plane that is about to take off, with the full knowledge that once it gets halfway down the runway it will not be able to stop in time, so that the pilot's only options are to either take off or crash. Scary stuff, though not as scary as the commitment shown by Australian motorcyclist Robbie Madison when he broke the motorcycle jump world record in Las Vegas on New Year's Eve 2007. This required hitting the launch ramp at nearly 100 mph, with no going back.

Limiting your own options is a strategy that I shall return to in chapter 6, but there is a much better strategy available for avoiding the sometimes terrible consequences of a game of Chicken: the two parties must find some way of coordinating their actions so that they can both escape from the situation with honour. This is exactly what Kennedy and Khrushchev eventually managed to do during the Cuban Missile Crisis: Khrushchev removed the missiles and Kennedy simultaneously lifted the blockade and later removed U.S. missiles from Turkey.

Coordination is always available to us so long as we can communicate. Indeed, communication is the key to negotiating coordinated strategies. Finding ways to communicate is the problem. It is a problem that becomes particularly serious when the game of Chicken involves many players.

The Volunteer's Dilemma

The Volunteer's Dilemma encapsulates all of those group situations in which the person making the first move risks losing out while others gain – but if nobody makes the first move, the loss can be devastating. Some examples that are commonly cited by game theorists include choosing who should be the one to jump out of a lifeboat to save it from sinking; deciding who should admit to being the culprit in a group offence so that all are not punished; and Yossarian's refusal to fly suicide missions in Joseph Heller's *Catch-22* ('What if everybody felt that way?' 'Then I'd certainly be a damn fool to feel any other way. Wouldn't I?')

The Yag·n Indians of Tierra del Fuego had a wonderful word for this: *mamihlapinatapai*, which means 'looking at each other hoping that the other will offer to do something that both parties desire to have done but are unwilling to do themselves'. It was described in the 1993 *Guinness Book of Records* as the 'most succinct' word in any language. It covers a multitude of situations, from siblings sorting out who is going to do the washing up or put the rubbish out to migrating wildebeest crossing crocodile-infested rivers.

Game theorists view the Volunteer's Dilemma as a multiperson (or multiwildebeest) version of the Prisoner's Dilemma, with multiple Nash equilibria to go with it (Box 3.4). I experienced a version of it in Australia when a bushfire started unexpectedly in the valley below my house. The fire was coming very fast, and the temptation was to rush out and start hosing the house down, hoping that one of my neighbours would take the time to phone the fire brigade before rushing out with *their* hoses. As it happened, several of us (including me) telephoned the fire brigade before we started hosing. (In fact, my wife was calling from an upstairs window that there were flames in the valley, and I called back, 'Yes, they're very pretty, aren't they?' 'Aren't you going to call the fire brigade?' 'Oh, yes [long pause]. I have already, actually.' She still hasn't quite forgiven me.) But what would have happened if we had all left it to someone else? With the intensity of the fire, we may all have lost our houses, instead of having four fire engines and two helicopters coming to our aid in time to protect us (we needed them all!)

> BOX 3.4

THE VOLUNTEER'S DILEMMA

The Volunteer's Dilemma, or *mamihlapinatapai*, is a game of Chicken with many participants. If just one person volunteers, everyone benefits *except* for the volunteer, but if no one volunteers, everyone loses out. The pay off matrix looks rather like that of the Tragedy of the Commons:

		EVERYONE ELSE	
		Someone else acts	**No one else acts**
ME	**Act**	**Benefit-Cost of action**	**Benefit-Cost of action**
	Don't Act	**Benefit without Cost**	**Big loss**

There is one vital difference, though: the cheater's Don't Act strategy is no longer dominant. It works if someone else volunteers, but it could be a disaster if no one else volunteers. That is the dilemma.

∎

Migrating wildebeest have a similar problem. When the herd comes to a river crossing with crocodiles waiting in anticipation, the animals that go into the water first don't have a great future. Those that come behind have a much better chance of making a safe crossing while the crocs are chewing on their bolder companions. But if none of them volunteers to go into the water first, the whole herd will be cut off from the pastures on the other side, and they will all starve. As with many human situations in which volunteers are required, the answer lies in a heavy hint. The animals that get eaten don't *want* to go in first. They stand on the bank looking at each other in nervous anticipation until pressure from those behind pushes them in. That's the hint.

When fear holds us back, it can be others who suffer. The fact that no one was willing to make the first move cost New York resident Kitty Genovese her life in 1964 when thirty-eight neighbours watched as she was stabbed to death in the courtyard of her Kew Gardens apartment. No one was willing to volunteer to risk injury or worse to themselves to save her. Indeed, being the volunteer can require a courage amounting to heroism. When a grenade was lobbed into the middle of a platoon led by Staff Sergeant Laszlo Rabel of the U.S. infantry during the Vietnam War, the platoon members would have died or been seriously injured if they had all stood back hoping that someone else would act. Staff Sergeant Rabel did act, throwing himself on the grenade and sacrificing his own life to save those of his companions.

The Volunteer's Dilemma is all around us, and it creates special pressures when the volunteer would actually be acting on behalf of others. Imagine that you are a resident of a drought-affected developing country and that food aid is being handed out from the back of a truck. Would you volunteer to stand back to let the food be handed out in an orderly and fair manner, or would you strive to get as much as possible for your starving family, regardless of fairness? That's the real-life problem of ma-mihlapinatapai.

Situations that require heroism and extreme self-sacrifice are fortunately rare. How can we go about choosing who should be the volunteer in less extreme circumstances? The problem is one of choice between two or more different Nash equilibria, in each of which a different party loses out while others gain. Game theorist William Poundstone reports an experiment that shows how difficult it can be to find the optimum solution and that also sheds light on human greed. The experiment was conducted by *Science 84* magazine, which published an article on cooperation and accompanied the article with an invitation to readers to send a card asking for either $20 or $100. The offer was for everyone to receive what they asked for, provided that no more than 20 per cent of the requests were for $100, otherwise no one would get anything. The editors ultimately chickened out of offering real money, but they would have been quite safe, since 35 per cent of those who responded asked for $100, in the hope that a sufficient number of readers would volunteer to ask only for $20!

In this particular case the participants were simply left guessing what others would do. When there is some sort of hint that leaves all participants realising that one Nash equilibrium is favoured over others, that particular equilibrium is called a *Schelling point*. Its inventor, the Nobel Prize-winning economist Thomas Schelling, described it as a 'focal point for each person's expectation of what the other expects him to expect to be expected to do'. The clue to a Schelling point may come as some sort of social convention, as when a man and woman are heading for the same door and the man politely stands back to let the woman go first, when bus passengers line up to enter a bus, or when people leaving a plane wait for those in the aisle to move ahead before leaving their seats. We also see this in conversations, in which the person talking may be thought of as being on the winning side of a Nash equilibrium; it may come in the form of a pause to let the other person take their turn on the winning side of the alternative Nash equilibrium.

Schelling points provide cooperative solutions to problems involving parties that would like to coordinate their actions but can't communicate to do so. Schelling's own example concerned two people who have to meet on a certain day in New York City but neither of them knows when or where. When he put the question to a group of students, the majority answer was 'noon at the information booth at Grand Central Station'. Its tradition as a meeting place made this a natural Schelling point. In a real-life example, two colleagues of mine managed to meet in Paris, even though neither of them could remember when or where to meet on the specified day. After trying the Eiffel Tower, one of them remembered that the other really liked churches, and they eventually met in the Notre Dame cathedral at 6:00 P.M.

Schelling points rely on implicit or explicit clues, and problems can arise when people give false clues. The former British prime minister Margaret Thatcher was famous for giving false clues during interviews. She would pause as though leaving space for the interviewer to ask another question and then, just as the interviewer opened his or her mouth, she would start up again without leaving room for the question to actually be uttered. According to psychologist Geoffrey Beattie, this may have arisen from her earlier speech training, which produced a 'drawl on the stressed syllable . . . and a falling intonation pattern associated with the end of a clause'. Both of these have been identified by other psychologists as turn-offering clues – that is, Schelling points.

I once tried an experiment in which I investigated just how long people would keep responding to false clues when I walked towards them on a crowded street, indicating that I was going to step one way but actually stepping the other way. My technique (always with a puzzled smile on my face, so as to avoid physical violence) was to watch what the other person was doing and then to step to the same side so as to block their progress. I kept on doing this until something happened to break the deadlock. My record, achieved in Tokyo (where I found that people are very polite), was seventeen successive sidesteps. My worst experience was in London, where after only three steps a gentleman in a pinstriped suit said, 'Could you please make your bloody mind up!' To counterbalance this, when I tried the experiment in Sydney outside a bar, the attractive young woman who was my unwitting subject said after a few moves, 'Well, if we can't get past each other, we may as well go in and have a drink.'

My experiment demonstrated the role of clues in social situations. But what if there are no clues? What strategy should we then adopt?

One answer is to forgo the assumption that the other party (or parties) is perfectly rational and to assume that they will sometimes make a mistake. Game theorists call this the 'trembling hand' assumption, and it can make your own choice of strategy easier by allowing you to eliminate those situations that are risky if you fear that the other party might make a mistake. When fire was racing towards my house, for example, the assumption that someone else would call the fire brigade would have been a very risky one if there had been a finite possibility that no one would be sensible enough to do this. This is why I made sure to make the call myself.

The Battle of the Sexes

Even when it's a choice between two good options, problems can arise, especially when each of those options involves a Nash equilibrium. Wendy and I face just such a problem when it comes to dividing our time between England and Australia. It's a problem that many people would love to have; we are able to chase the sun, spending spring and summer in England before moving to Australia to spend spring and summer there, with lots of friends in both places. 'Lucky devils,' you might say, and you would be right. But we still have a problem, represented by the diagram in Box 3.5.

The problem is that Wendy, being English, would really prefer to live in England for most of the time and to visit Australia occasionally. I, on the other hand, would prefer to live in Australia, where I was born, with occasional visits to England. Neither of us, however, would like to live apart – living together in either Australia or England would be preferable to that. Each of these possibilities is a Nash equilibrium – but how can we choose between them? What is the best solution?

We found that we were trapped in one of the most enigmatic, irritating, and puzzling dilemmas that game theorists have discovered. It is called the Battle of the Sexes – not because it has anything to do with man versus woman, but because the first example to be looked at concerned a man who wanted to go to a baseball game while his wife preferred the movies. A better, if less evocative, title might be *Unfair or Inefficient*, because these were the two choices that we faced, depending on which game theory solution we chose.

Our first choice (the one that we used instinctively until we came across a better solution) was simply to take turns, living

➤ BOX 3.5

THE BATTLE OF THE SEXES

The Battle of the Sexes is ill named, because it is not really a battle but more a matter of choosing between two quite reasonable Nash equilibria in such a way that both participants end up in the same one. It's a bit like Chicken without the disaster scenario, as the following diagram shows for the game theorists' favourite example: one person wants to go to a ball game and the other would prefer to go to a movie:

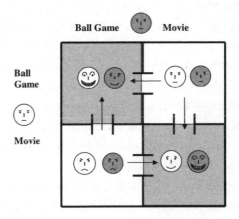

One important difference between the Battle of the Sexes and Chicken is that the two Nash equilibria are on the diagonal running from top left to bottom right, which means that they correspond to the same rather than opposite choices. The problem that the participants face is which to choose. If they can communicate, they can flip a coin, and neither will have an incentive to cheat on the resultant agreement. (This sort of randomisation is a great example of a mixed strategy.) If they cannot communicate, each must guess what the other is most likely to do.

for a bit more than half the year in England and for the remaining time in Australia. We didn't divide the times to be completely equal, because I gain some benefit from living in England, where I do most of my writing and broadcasting. Even so, the division was unfair, for no matter where we were, one of us would have preferred to be in the other place!

One solution was to draw up a list of pluses and minuses for each of us and try to balance it out – a sort of cost-benefit analysis to produce the optimum balance between individual benefit and marital bliss. This is just what Charles Darwin did when he drew up a list of pros and cons before deciding whether or not to propose marriage to his cousin Emma. Among the benefits of marriage that he saw were: 'Object to be beloved and played with. Better than a dog anyhow; home, and someone to take care of house; charms of music and female chit-chat; and a nice soft wife on a sofa with good fire and books and music perhaps.' As against these, the bachelor life offered 'Conversation with clever men at clubs; not forced to visit relatives and bend in every trifle; [absence of] anxiety and responsibility; and money for books.' The clincher, though, was: 'My God, it is intolerable to think of spending one's whole life, like a neuter bee, working, working, and nothing after all – No, no, won't do.' His conclusion was 'Marry, Marry, Marry Q.E.D.' It is a conclusion that I would thoroughly endorse after twenty years of marriage to my present wife, although not necessarily for all of Darwin's reasons.

The trouble with using this sort of cost-benefit analysis to decide between two Nash equilibria is that the solution isn't efficient. In our case, we decided that I would come out to Australia several weeks before Wendy and go back to England a couple of weeks later, sacrificing some of our time together so that we could each spend more time in our favoured place. I managed to get a fair bit of writing done but did not enjoy being on my own. Wendy, on the other hand, was stuck with cleaning up our house in England before joining me in Australia. Our mixed-strategy solution trapped us in a Nash equilibrium, but it was an equilibrium that was not as favourable to either of us as what either of the 'pure' strategies would have led to.

There was an answer, though, discovered by the Israeli-American game theorist Robert Aumann, who shared the 2005 Nobel Memorial Prize in Economics ,for having enhanced our understanding of conflict and cooperation through game-theory analysis'. Aumann's answer to the Battle of the Sexes dilemma was for both people to agree to some random way of determining their strategy, such as tossing a coin or drawing a card. In our case it was the toss of a coin, with the prearrangement that if it came up heads I was to stay longer in England before coming out with Wendy, while if it was tails she was to come out to Australia earlier with me.

We were both better off with this arrangement. Aumann called it a 'correlated equilibrium', because it binds the choices of the two parties together in a very neat way. It may seem trivial when a coin toss decides the issue, but Aumann had actually come up with a solution concept that can be more powerful than the Nash equilibrium, and that can even help to resolve some games of Chicken in which the participants seem to be locked into a mutually destructive collision course and neither is prepared to give way. The logic of self-interest that locks them in can, with Aumann's unusual twist, help to undo the lock. It all depends on agreeing to a rule for randomising their choice of strategies and then finding a disinterested third party who is willing to apply the rule and tell each party privately what it tells them to do, without telling them how the rule has affected the other party. It is simple in principle but sometimes tricky in practice.

Stag Hunt

Finally, there is Stag Hunt (Box 3.6), which game theorist Brian Skyrms believes to be more relevant to the problem of social cooperation than the Prisoner's Dilemma. The name comes from a story told by the French philosopher Jean-Jacques Rousseau about a group of villagers hunting a deer: 'If a deer was to be taken, every one saw that, in order to succeed, he must abide faithfully by his post: but if a hare happened to come within the reach of any one of them, it is not to be doubted that he pursued it without scruple, and, having seized his prey, cared very little, if by so doing he caused his companions to miss theirs.'

Rousseau saw the story as a metaphor for the eternal tension between social cooperation and individual freedom. In his words (referring to the social contract between the individual and the state), 'True freedom consists in giving up some of our freedoms so that we may have freedom.' This is exactly what the people in a stag hunt do when they give up their freedom to make the certain catch of a hare in order to cooperate in the pursuit of a bigger but less certain goal of catching a stag. Skyrms draws an intriguing parallel with the way that many societies (especially democratic ones) operate: 'The problem of instituting, or improving, the social contract can be thought of as the problem of moving from the riskless Hare Hunt equilibrium to the risky but rewarding Stag Hunt equilibrium.'

> BOX 3.6

STAG HUNT

Stag Hunt is like an inverted Prisoner's Dilemma: the Nash equilibrium and winning position is 'cooperate, cooperate' rather than 'cheat, cheat' for all parties. It sounds ideal, but let's put our participants Bernard and Frank out hunting in the wild and see what might happen in practice:

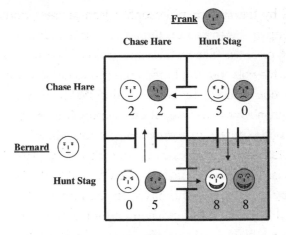

There is a very clear Nash equilibrium at the bottom right, with no temptation for either to cheat – unless he believes that the other may cheat. If one cheats, the best option for the other is to cheat as well. The multiperson version of this dilemma needs no further amplification, except to say that it is a very common one.

The choice of strategies in Stag Hunt seems at first glance to be a no-brainer. The reward is greater for cooperating than it is for cheating (in game theory parlance, defecting), and so we should always cooperate and reap the greater reward. This is the dead opposite of the Prisoner's Dilemma, in which the dilemma comes from the fact that the reward to the individual is always greater for cheating, no matter what the other party does. The fly in the ointment with Stag Hunt, though, is the element of risk.

The Prisoner's Dilemma is pay off dominant. In other words, the reward is what matters most, and you choose your strategy to maximise that reward. Stag Hunt, however, is risk dominant, which means that the favoured Nash equilibrium is the one that entails the least risk.

While I was writing this section, I saw an interesting example of risk-dominant strategies in the film *Amazing Grace*, which tells the story of the English politician William Wilberforce and his fight for the abolition of slavery. Many other politicians might have supported his stance and brought slavery to an end much more quickly were it not for the fact that those whose constituents profited from the slave trade were afraid to speak up unless a sufficient number of others did. Their voting strategies were risk dominant – that is, they were designed to minimise the risk to their political careers.

William Poundstone gives a more recent example in his book *Prisoner's Dilemma*. U.S. senators were voting on President George H. W. Bush's 1989 proposed constitutional amendment to make the burning of the American flag a federal crime. 'Most opponents of the bill objected to it as a violation of freedom of expression,' Poundstone says. 'At the same time, they feared that if they voted against it . . . their opponents would brand them as unpatriotic.'

Stag Hunt represents the fragile circumstances in which so many of the world's people now live, especially when it comes to the preservation of individual liberties, freedom of expression, and even the freedom to hold private conversations. For example, when I visited Tibet recently I found that it was impossible to talk freely with individual Tibetans about the problems in their country because they were frightened that their conversations, or even the fact that they had had a conversation with a Westerner, would be reported by one of their neighbours to the authorities. The Stag was the freedom to talk. The Hare was the more certain reward of spying and reporting secretly on your neighbour. The Divide and Rule strategy works because it is the risk-dominant strategy in a Stag Hunt scenario. It is not an easy thing to change, even with the tools of game theory. As Skyrms says: 'The news from the frontiers of game theory is rather pessimistic about the transition from hare hunting to stag hunting. . . . For the Hare Hunters to decide to be Stag Hunters, each must *change her beliefs* about what the other will do. But rational choice-based game theory as usually conceived, has nothing to say about how or why such a change of mind might take place.'

The real issue is not just getting individuals to change their beliefs about others; it is getting a whole group of people to do this in a coordinated manner. That's just the first step, though. The next step is persuading people to stick to their new positions and not change their minds again. This is the second fundamental problem of cooperation. In the following chapters I investigate just how cooperation might be achieved in the face of the various temptations to cheat.

4

Rock, Paper, Scissors

THE FIRST SELF-ENFORCING STRATEGY I investigated comes from the childhood game of Rock, Paper, Scissors, a worldwide game whose name varies from country to country. One of my favourites, from Japan, is Chief of the Village-Tiger-Mother of the Chief of the Village. Others include Snake-Frog-Slug (Japan), Elephant-Human Being-Earwig (Indonesia), Bear-Man-Gun (Canada), and Bear-Cowboy-Ninja (may be specific to Milwaukee!)

Whatever its name, children aren't the only ones to play it. Adults have also been known to use it when they cannot agree or would rather leave a decision to chance. George Washington is reputed to have played it with Lord Cornwallis and the Comte de Rochambeau to decide who would be the last to leave Cornwallis's tent after the signing of the British surrender at Yorktown in 1781. (The story goes that Rochambeau won, which is why the game is still called Ro-Sham-Bo in some quarters.) More recently, a Florida judge ordered two attorneys to play it when they could not agree on where to hold a deposition, even though their offices were just four floors apart in the same building!

In the case of the attorneys, it needed an external authority to enforce the decision, but game theorists have discovered that the introduction of a third player totally changes the nature of the game. Now there is no need for an external authority, because there is an inbuilt balance among the strategies of the three participants, and no one strategy can dominate. Nature uses such a balance to maintain a diversity of species with different strategies for survival. Game theorists have shown that we can use a similar balance to limit the number of cheats in the Free Rider dilemma. Here I examine how the balance arises and how we might be able to use it in practice.

I started with the two-player game. Most of us know the simple rules. The players hold their right hands out simultaneously at an agreed signal to represent a rock (closed fist), a piece of paper (open palm), or a pair of scissors (first and second fingers held apart). If the two symbols are the same, it's a draw. Otherwise rock blunts scissors, paper wraps rock, and scissors cut paper, so the respective winners for these three outcomes are rock, paper, and scissors. It's dead easy.

Rock, Paper, Scissors is a zero-sum game. If you count a win as +1 point, a loss as −1 point, and a draw as zero, for example, the sum of the wins, losses, and draws for a game is always zero. To a game theorist this means just one thing – the optimum strategy can be worked out from the Minimax principle. This leads to the intuitively obvious conclusion that the best approach, in the absence of any information about the opponent's intentions, is to use a mixed strategy, throwing rock, paper, and scissors at random, with equal probability for each. When both players play in this way, they have an equal chance of winning, losing, or ending up in a tied game.

The psychological appeal of the game is that both sides feel that they are in control because they have a choice of moves. This means that if they can work out what their opponent is likely to do, they can always pick a play that will beat it. This is the position that the famous auction houses of Christie's and Sotheby's found themselves in when Takashi Hashiyama, chairman of the Japanese Maspro Denkoh Corporation, decided to auction his firm's collection of Impressionist paintings. He asked the auction houses to decide between themselves who would get to sell the valuable collection, and in an e-mail he 'suggested to use such methods as rock, paper, scissors'. The two firms had the weekend to decide which move to play, but their approaches to the problem were very different. Sotheby's claimed that they 'didn't really give it that much thought' and went with paper, presumably in the hope that Christie's would play rock. Christie's won out, though, when they took the expert advice of art director Nicholas Maclean's eleven-year-old twin daughters, Flora and Alice, who suggested scissors because, as Alice explained, 'everybody expects you to choose rock'.

Maybe the girls had learned from the episode of *The Simpsons* in which Bart thinks to himself, 'Good ol' rock. Nuthin' beats that!' which Lisa, of course, predicts. In fact, Sotheby's was unlucky, because they had chosen the statistically best strategy. Scissors tends to be played slightly less often in tournaments (29.6 percent of the time, compared to the statistical expectation of 33.3 percent), which means that it is worth playing paper slightly more often. With that minor modification aside, complete randomisation is the best option.*

The Sotheby's approach was perfectly reasonable, because they had no way of knowing what the Christie's team was going to do. If they had, they could certainly have chosen a strategy to beat it. This no longer applies if a third player is introduced, as I discovered when I introduced the game to my five-year-old grandson. His mother and I demonstrated the game to him, and he thought about it for a while before proudly announcing, 'I always play rock!' This made our strategy rather obvious, so long as just one of us was playing against him. But when the three of us tried playing simultaneously, something very peculiar happened. His preannouncement of a strategy meant that his mother and I could never beat each other without letting him win against one or the other of us. If I played paper, for example, she could beat me by playing scissors, but only at the expense of losing to his rock.

The balanced tension among the three possible outcomes arose because of what mathematicians would call the *intransitive* nature of Rock, Paper, Scissors; in other words, the fact that rock beats scissors and scissors beats paper does not imply that rock beats paper. Instead, the three strategies are locked in an endless circle because paper beats rock.

---

*The only problem with complete randomisation is that it is difficult to achieve in practice, and most people end up following some sort of pattern, which a good opponent may be able to predict. To overcome this, I worked out a way of randomising strategies that no opponent could predict, because I could not predict it myself from one game to the next. When I tested my strategy against an online computer program, it proved to be remarkably successful. (See the box at the end of this chapter.)

A similar balance among intransitive strategies is used in nature to produce a balance among species that use different reproductive strategies. The Californian side-blotched lizard provides one interesting example. There are three types of males: those with orange throats, yellow throats, and blue throats. Those with orange throats adopt an aggressive strategy and defend large territories with many females. Those with yellow throats can beat this by adopting a sneaky strategy that allows them to mate with some of the females in the orange-throated male's harem when he isn't looking. The strategy of the yellow-throats, though, is beaten by blue-throated males, which keep small harems that they guard jealously to keep strangers out. But the blue-throated males are beaten in turn by the aggression of those with orange throats in a beautiful symmetry of strategies that is identical to the symmetry of the strategies in Rock, Paper, Scissors.

It wouldn't do any of the three types any good to give up their own strategy and use one of the others. If the orange-throats, for example, switched to the sneaky strategy of the yellow-throats, they could never beat the blue-throats, who would soon become dominant. Nor could the orange-throats switch to the blue-throat strategy, because this would simply mean that the yellow-throats could themselves switch to the aggressive orange-throat strategy and quickly become dominant. All three are doing the best that they can, given the strategies of the other two. Evolution, in other words, has produced an optimum, balanced set of best responses for each type of lizard in response to the best strategies of the other two types, and the net result is that the population of each type of male varies over time but maintains an average of one-third of the total, which is the best that any of them can do.

The natural balance produced by the Rock, Paper, Scissors scenario is not confined to lizards. Researchers from Stanford and Yale have discovered that the same scenario is responsible for preserving biodiversity in bacterial neighbourhoods. The bacteria concerned are *Escherichia coli* – the type that are found in all of our digestive systems. The researchers mixed three natural populations together in a petri dish. One population produced a natural antibiotic called 'colicin' but was immune to its effects, as snakes are immune to poisoning by their own venom. A second population was sensitive to the colicin but could grow faster than the third population, which was resistant to colicin. The net result was that each established its own territory in the petri dish. The colicin producers could kill off any nearby bacteria that were sensitive to the colicin, the colicin-sensitive bacteria could use their faster growth rate to displace the colicin-resistant bacteria, and the resistant bacteria could in turn use their immunity to displace the colicin producers!

This type of self-enforcing balance among several different strategies has been shown to be an important component of biodiversity. If just one species is lost, together with the strategy that it uses to survive, the balance of strategies with the others is also lost, to the detriment of all except one. If orange-throated male side-blotched lizards suddenly disappeared, for example, the yellow-throats would soon follow, since their sneaky strategy is beaten by the blue-throats' defensive strategy. Only the blue-throats would remain. The same damaging process happens in plant communities: when one species is lost, others can soon follow. The Rock, Paper, Scissors scenario, in which each strategy is the best response to the other two, maintains the balance.

Opting Out as a Third Strategy

The Rock, Paper, Scissors scenario in nature creates a situation in which no one strategy can ever dominate the others. Game theorists have argued that a similar approach could be used to help us to resolve the Free Rider problem, in which the cheating strategy of taking advantage of a communal resource without contributing to its upkeep can rapidly come to dominate the cooperative, contributing strategy in the absence of constraints such as social disapproval. I saw one example of how cheating can quickly become dominant in our local preschool when a new child arrived who had never been required to put his toys away after playing with them at home – his indulgent parents had always done it for him. When the time came for children to cooperate by putting their toys away at the end of the day, he rebelled and cheated by continuing to play with them. Soon other children were following his example – if he could keep playing, why couldn't they? The teacher was too weak to stop them, and when the parents started arriving to pick up their children, the whole kindergarten was in an uproar and there were toys everywhere.

Could the teacher have done anything about it, without re-sorting to the forbidden use of force? Game theory suggests that she could have, if she had adopted the Rock, Paper, Scissors an-swer of introducing a third strategy that could beat the cheating strategy, but be beaten by the cooperative strategy. One thing she might have done would have been to tempt the children to leave the toys entirely (by offering them ice cream, say), with the proviso that children who first put their toys away would receive an additional reward (perhaps more ice cream).

Would it have worked? All I can say is that it did·when I tried it. I was cast in the ignominious role of a clown at a children's party, with the difficult task of getting the children to tidy up at the end when they wanted to continue playing. I offered special chocolate-coated ice cream to all who were willing to stop playing, with the additional reward of a chance to hit me with a water balloon if they tidied up first. Some children chose to keep playing with their toys. Others took the ice cream but weren't interested in tidying up. Most, though, tidied up in return for the chance to give me a soaking.

By giving the children a chance to opt out of the keep-playing (cheat) and tidy-up (cooperate) choices by offering a third alternative, I had set up a fresh balance between these two options – just as my grandson did when his permanent rock strategy in our three-way Rock, Paper, Scissors game forced his mother and me into a situation where neither of us could ever beat the other. Game theorists call such a third strategy a Loner or Volunteer strategy. The effect of introducing such a strategy (which might also be called 'opting out') has been described in the following way: 'volunteering relaxes the social dilemma: instead of defectors [the game theorist's word for cheats] winning the world, coexistence among cooperators, defectors and loners is expected'.

In other words, opting out can actually promote cooperation among those who have chosen not to opt out! This is exactly what happened in my English village when we decided to get together to bring a group of children to our community from the stricken Chernobyl region for a recuperative holiday. More and more people opted out of doing the actual organisation, forcing the rest of us to cooperate more closely to make it happen. Among our committee of cooperators there were inevitably one or two who rested on their oars and left the work to others, so that in the end we had a dynamic balance of cooperators, defectors, and loners – just as game theory would have predicted.

Manfred Milinski and his group at the Max Planck Institute of Limnology in Germany have studied how this balance comes about in formal laboratory experiments. *Limnology* sounds like the study of arms and legs, but it is in fact the study of freshwater lakes and ponds. Milinski and the other members of his group are evolutionary ecologists who are interested in the evolution of cooperation in communities of freshwater organisms. The organisms that they chose to study on this occasion, though, were unlikely to have been very interested in fresh water, since they were first-year biology students who were probably happier when pickled in alcohol.

The pickle that the experimenters put them into involved the decision of whether to be a cooperator, a free rider, or a loner in a computer game that rewarded them in hard cash according to the choice they made. Loners received a small payment if they chose not to join the group and not play the cooperation game. They could earn a bigger reward if they volunteered to join the group and participate as a cooperator, and an even bigger reward if they volunteered to join the group and then chose to cheat on the cooperation and become a free rider. The sting in the tail was that if too many of them chose the free rider option, the rewards for cooperating and free riding would both drop to the point where they would have been better off remaining loners.

You can see the similarity to the Rock, Paper, Scissors scenario in that as soon as one strategy was used by too many participants, it could be beaten by one of the others. The overall result was the same as if the students had been Californian side-blotched lizards, with the populations of cooperators, free riders, and loners oscillating around a mean value of one-third of the total population each. Free riders could exploit a large group of cooperators, but if there were too many free riders, it was better to be a loner. When there were lots of loners, however, the size of the remaining group became smaller, to the point where individuals could no longer benefit by becoming free riders, just as each member of a tug-of-war team would lose more than they gained by relaxing and not putting in their full effort while the result was in the balance.

Milinski's results showed *why* they were able to claim that the strategy of volunteering 'relaxes the social dilemma'. It is because 'when loners are the most frequent the public group size is reduced, which invites cooperation because the game is no longer a dilemma in small groups'. In other words, it no longer pays to cheat by taking a free ride.

I saw an example of the small-group effect when I lived in a small English village and my elderly neighbour's house was burgled and several valuable clocks were stolen. The burglar was boasting about his exploit to his mates in the local pub and was told by them to take the clocks straight back! Because it was a small community, it was not worth their while to have any one of them known as a thief.

In summary, 'volunteering [that is, opting out] does not produce overwhelming cooperation, but it might help to avoid the fate of mutual defection in many human collective enterprises [by reducing the dominance of the defection strategy]'.

Opting out is only one possible third strategy, however. There are other approaches to creating a balance when it comes to a competitive triangle. One of these is to look at it as a *truel*.

The Truel

A truel is similar to a duel, except that three parties are involved. The presence of the third party can create paradoxical situations that have great relevance to many situations in real life.

One of those paradoxes is illustrated by the hypothetical case of three male logicians who get into an argument over the finer points in game theory. The argument becomes so violent that they decide, being male, that it can only be settled by a shoot-out to leave the best man standing. Being logicians, though, they have to work out some rules, and they come to the conclusion that the fairest thing is for the worst shot to go first, followed by the second-worst shot, followed by the best shot, and so on in sequence until only one is left standing. Statistically, the worst shot only hits his target one-third of the time, while the second-worst shot hits it two-thirds of the time, and the best shot never misses. If you were the worst shot, whom would you aim at?

The answer is that you should fire into the air! If you fire at the second-worst shot and hit your target, you are a dead man. If you fire at the best shot and kill him, you have only a one-third chance of survival. In other words, by initially killing one of your opponents you would only make your chances worse, because the remaining opponent would then shoot at you instead of the third man. By missing, you get another shot, with better odds.

There are many actual parallels to this imaginary scenario. One occurs in the world of chess and bridge tournaments, many of which are played on a Swiss format, with losers from the first round playing other losers from the first round, and so on. When I used to play in such tournaments, I soon worked out that the best strategy was to be sure to lose in the first round, so as to play weaker opponents thereafter.

I found that standing back to let the strong ones fight it out before entering the fray worked in many areas of life. This applied especially to committee meetings. Rather than enter a debate at an early stage, I could frequently get my way simply by waiting until others had argued their points vociferously to the point of exhaustion, and then bringing up my own point at the last minute.

Marc Kilgour and Steven Brams, the scholars who first analysed the truel, have pointed out several fascinating examples. One well-known one was the truel fought in 1992 among the three major U.S. television networks over anchors and formats for their late-night shows. ABC effectively fired into the air by sticking with its popular *Nightline*. This forced CBS and NBC into a duel over which comedian, David Letterman or Jay Leno, to attempt to hire in order to capture the late-night TV entertainment audience.

A more serious example was the extended nuclear deterrence during the Cold War, in which the participants were the United States, Western Europe, and the Soviet Union. A duel between Western Europe and the Soviet Union might have developed as a full-scale war. The presence of the United States, with an implied threat of nuclear retaliation against the Soviet Union if it, say, invaded West Germany, turned it into an extremely dangerous truel.

Such conflicts are, of course, usually too complex simply to be analysed as a truel. Kilgour and Brams argue that we can still learn lessons, so long as we recognise that we need carefully to identify the rules under which the real-life versions are being played out. This is particularly important because optimal play can be very sensitive to a slight change in the circumstances. One of the most robust lessons is that the strongest participant is often in the weakest position, since it will be an early target. As a corollary, they argue, '[in] contemplating the consequences of a long and drawn-out conflict, truelists may come to realize that their own actions, while immediately beneficial, may trigger forces that ultimately lead to their own destruction'. The history of attempts by one strong country to suppress rebellion and terrorism that are supported by other strong countries makes the point, from the American Revolutionary War to the present-day conflicts in Afghanistan and Iraq.

Another real-life problem pointed out by Kilgour and Brams is the fragility of any pacts that might be entered into by the parties. This applies particularly in politics. To give one example from my own country, the Australian state of Tasmania was left without a government after a truel – involving two major political parties and the small Green Party that held the balance of power – when repeated attempts by both major parties to form an alliance with the Green Party failed. Another example comes from Italy, where both houses of Parliament have had to be dissolved seven times since the Second World War because of the failure to form a stable coalition government.

How can we cooperate to form more stable pacts and alliances? It means finding ways to make breaking the pact a losing proposition for all sides. In the next chapter I investigate whether and how this might be achieved.

➤ HOW TO HOLD YOUR OWN AT ROCK, PAPER, SCISSORS

I say 'hold your own' advisedly, because I wouldn't back myself to improve on the odds by predicting someone else's strategy. If you can do it, good luck to you. My main concern here is to find a way of stopping others from beating *me* more than half the time, and the answer is to find a truly unpredictable randomising strategy and stick to that.

There are many possibilities. My own approach was simply to memorise the first twenty or so digits in a transcendental number – such as e or π – because I knew that there was no way to predict the next number in such a sequence from the ones before. I then played rock if the number was 1, 2, or 3; paper if it was 4, 5, or 6; and scissors if it was 7, 8, or 9. On 0 I would play paper (for a reason that I give below) or just go with a whim. For π, for example (3.14159265358979323846 . . .), the sequence of moves would be rock, rock, paper, rock, paper, scissors, rock, paper, paper, rock, paper, scissors, scissors . . .

Such sequences are as close to random as one can get, so knowing what one digit (or throw) has been gives no clue as to what the next one will be. You can also start at different places in the sequence for a different game, or even run it backwards if you have a good memory for numbers (which is not as hard as it looks).

I tried it out against Roshambot – an artificial intelligence algorithm designed by computer scientist Perry Friedman, who is now a professional poker player in Las Vegas. Friedman told me (*after* I had played against the program) that it 'looks for patterns in your play. The bot looks for move pair matches and weights them by the length of the string of matches, so if it sees a pattern that matches the last five pairs of moves, it weighs that more than a pattern that matches just the last three pairs'.

➤

He also claims that 'if you are truly random, it can't dominate, or be dominated. If you use true random play, any result is truly random and represents just that, random fluctuation'.

Judge for yourself from these results over five hundred games, obtained by matching moves to digits in the transcendental number e, where 0 = paper, 1–3 = rock, 4–6 = paper, and 7–9 = scissors:

Win	Lose	Draw
185	159	156

It looks as though the program outsmarted itself by looking for patterns where there were none, although Friedman argues my success was mainly due to chance.

To outsmart most human players, you don't need to look for patterns – in fact, it's better not to, because it can take quite a sophisticated algorithm to find them when they are there. It's better to rely on statistics, which show that most players have a slight tendency to play rock more than they ought, and less of the others. In 1998 Japanese mathematician Mitsui Yoshizawa studied throws from 725 people and found that they threw rock 35 per cent of the time, paper 33 per cent of the time, and scissors 31 percent. People playing the online Facebook game Roshambull come in at rock 36 per cent, paper 30 per cent, and scissors 34 per cent. So you should play paper and scissors slightly more often than you play rock, but how you draw the balance between the first two depends on whether you are playing a live player or not, or perhaps whether you are playing in Japan or on the Internet!

5

Let's Get Together

COMMUNICATION AND NEGOTIATION are the twin keys that can unlock social dilemmas. They let us share information and ideas, form alliances, and agree on joint, coordinated strategies. Unfortunately, information can be false, ideas can be misleading, and alliances can pit the strong against the weak or break up and rearrange themselves like soap bubbles in a bath. How can we make communication more reliable, negotiation fairer, and alliances more stable?

Communication

Animals have developed many ways to send unambiguous messages. Herrings, for example, communicate by farting. Their farts create a clicking noise called a 'fast repetitive tick' that other herrings can sense but that predators seem to be unable to detect. This allows the herrings to keep their positions so that the shoal can move in unison, even at night when the fish can't see each other.

Among other organisms, small boys in a classroom come closest to using this method of communication. My Australian compatriot Clive James, the author and broadcaster, used to amuse his schoolfellows by interposing a noisy, gaseous commentary on his teachers' explanations, in the manner of the famous *Moulin Rouge* performer Le Pétomane, who had sufficient control to be able to fart *La Marseillaise*.

Farting, however, can communicate only limited information. A better method might be that used by bees, who communicate by dancing. A bee that has found a fresh source of nectar will return to the hive and perform a complex 'waggle dance' in front of the other bees to tell them what direction and how far they will have to fly to collect the nectar. Ants, on the other hand, simply lay down an odour trail for the other ants to follow to the food source, which is why we see ants travelling in long columns that seem to be magically coordinated.

Farting, dancing, and laying down odour trails all have their parallels in human communication, but the closest approach in nature to our own methods comes from the humpback whale. Male humpbacks produce songs that have a hierarchical syntax (a grammar and a structure), just as human language does. The content of these songs, which can last for up to thirty minutes, has been analysed by scientists from Harvard University and the Woods Hole Oceanographic Institute. The meaning of the songs is still not clear. (To paraphrase researcher Peter Tyack, they are not reciting *Hamlet*, but they could be singing love songs.) But the whales' language is certainly being used to communicate specific information to other whales, sometimes halfway around the globe.

The language used by whales only permits them to transmit information at a leisurely rate of just one bit per second, where a bit is the smallest piece of information that allows a distinction between two possibilities. At first glance, our own language is not very much faster. Even President Kennedy, who holds the world record of 327 words per minute, was still speaking at only 16 bits per second (for comparison, a slow computer modem still transmits information at a staggering 56,000 bits per second). My own more modest speaking rate of 200 words per minute lets me communicate at 10 bits per second, only ten times faster than a whale.

The difference between my communication and that of a whale is that I put my bits together in distinguishable sets (known as *phonemes*), at an average rate of 5.5 bits per phoneme, which I then further unite to form words (at an average of 4 to 6 *phonemes* per word). These words can be combined in millions of different ways to produce a complex language that is rich in meaning. It is this complexity that permits me to use language not just to communicate, but to *negotiate*.

Negotiation

As we know, many animals use ritualised displays to negotiate for mates, food, and territory. Humans also use body language and ostentatious display. A flashy car can be seen as the equivalent of a peacock's tail, and a warning frown is our equivalent to the baboon's display of a colourful backside. Fortunately we have no equivalent for what happens when two male hippopotamuses get into a confrontation and (to quote the *Ultimate Irrelevant Encyclopaedia*) 'the one who produces the most excrement is usually the eventual winner, boosted to victory by the smell of his own dung. If that fails, they are wont to confound the enemy with foul-smelling belches'.

Animal displays and responses are genetically programmed, and they lead to predictable outcomes. Sometimes those outcomes include violence, just as ritualised threats between rival groups of drunken football supporters can lead to violence. I sometimes think that such supporters are genetically programmed themselves, but on the one occasion when I was confronted by a group of them, I discovered that the use of language allowed a flexibility of negotiation (a key tool in resolving social dilemmas), which helped me to produce a non-violent outcome.

I was travelling with a physicist friend in a train that filled up at one station with inebriated football supporters whose team had just lost an important game, and they were looking for trouble. It nearly came our way when my friend showed me an experiment that involved holding two fingers up to the light, only to find that he was holding them right in front of the face of one of the more vociferous drunks. To say that the drunk was annoyed by the gesture, which in Australia has a very rude connotation, would be putting it mildly. I hastily explained that it was a scientific experiment, and when he showed a flicker of interest I took the opportunity to show him how a dark band appears between the closely spaced fingers and praised him for being able to spot the effect so quickly when he tried it out for himself. He proudly showed off his discovery to his mates, and when we left the train it was full of intrigued drunks holding two fingers up to the light.

Without language, we could have been in real trouble. With it, I was able to explain the meaning of a gesture that had been misinterpreted. The flexibility of language also allowed me to offer a better reward to our train-mate than the pleasure of hit-

ting my friend, in the form of the kudos that he gained by showing off his new discovery to *his* friends.

My only alternative in this situation would have been to
make a threat, which I was hardly in a position to do, and
which simply would have escalated the situation. When the late
George Melly, the jazz singer, found himself in a similar situation, however, he discovered a threat that was *really* effective.
He was confronted by a gang of drunken youths after a concert
and, not being able to think of anything else to do, he pulled
a book of Zen poetry from his pocket and started to read the
weird-sounding verses aloud. The startled teens ran off, convinced that he was mad and frightened by the thought of what
he might do next.

Threats and promises are the twin tools of negotiation. The
choice of which to use, however, depends very much on circumstances. To be effective, they must be believed. A parent
screaming, 'I'll kill you if you don't stop that' at a child is unlikely to be believed. 'I'll take your ice-cream cone away" or 'I'll
buy you an ice-cream cone' is likely to be much more effective.

Threats are cheaper than promises, because if a threat is effective, it will not need to be followed through. The promise of
a reward (where possible), however, can be less likely to lead to
escalation, which is always a possibility with threats if they are
seen to be hollow. That's not to say that promises can always
avoid escalation – promising a reward to a blackmailer is often
the first step in an escalating series of demands that eventually
bleed the victim dry, and corrupt officials tend to become ever
greedier in the bribes that they demand. Rewards, though, are
still the preferred option for most everyday negotiations. Shopping, for example, is effectively a negotiation where we promise
a reward by saying, 'I will give you this money if you will give

me those goods.' The shopkeeper is doing the same thing in reverse, saying, 'I will give you these goods if you give me that money.' Sometimes they even promise reward points as an additional incentive to get you to hand over your money.

This may seem like an unnecessarily roundabout way of thinking about something as simple as shopping, but my wife and I found just how helpful it could be when we bargained for some clothes on our first visit to India. We agreed on a price of 300 rupees (about £4) with the shopkeeper and handed over a 500-rupee note, expecting the clothes and 200 rupees' change. But we didn't know India! Instead of giving us the change, the merchant wanted to sell us more clothes in lieu, and he was prepared to bargain all day rather than give us the actual change. We have now learned our lesson, and we play the shopkeepers at their own game by going armed with lots of small notes and at first handing over slightly *less* than the amount we originally agreed on, then promising to give the merchant some of the clothes or other goods back when they complain. What we are doing is trying to reach a position where the agreement is a genuine one, with no room for further manoeuvre. We feel no compunction about using this approach, because our Indian friends assure us that in almost every case the merchants know exactly how much their bottom line is, and they usually succeed in getting a high price from us anyway.

Coalitions

Game theorists would say that we were trying to form a coalition with the shopkeeper in India. Most people think of coalitions in terms of political parties, or of nations with joint (usually warlike) objectives. Game theorists have extended the term to mean any alliance in which the members coordinate their strategies to work cooperatively toward a common objective. In the eyes of a game theorist, marriage is a coalition (though not always a very successful one). So is a sports team. So are two pedestrians stepping aside to get past each other, or a shopper and merchant exchanging money for goods, because in both cases they must form a temporary alliance to coordinate their strategies. Just to complete the set, an individual who fails to form alliances suffers the indignity of being called a '*singleton* coalition'! In the game theorist's eyes there's just no getting away from coalitions.

Negotiation to form a cooperative coalition is what my editor and I have done in preparing this book. My strategy has been to understand and work out the ideas for myself and to present them in logical order with interesting examples to help the reader understand them. Her strategy has been to gently help me focus on what they might mean for a broader audience, both in everyday life and in the context of the worrying problems that the world faces. Many books have been written about strategies for conducting such negotiation processes, in contexts ranging from politics and international diplomacy to business management, the running of organisations, and personal relationships. It is not my purpose to add to this list, even if I could. What I am interested in is where the process leads and what we must aim for if it is to produce successful cooperation.

One obvious objective is to establish coalitions that make it possible for all parties to coordinate their strategies and trust each other to stick to the agreed strategy. This allows them to escape from social dilemmas and discover cooperative win-win solutions instead. According to game theorist Roger McCain, this is always possible because '**if people can arrive at a cooperative solution, any non-constant sum game can, in principle,**

be converted to a win-win game.' (The emphasis is mine.) If it were physically possible, I would have designed this book so that the statement could jump out of the page and run around screaming its message, just as it jumped out at me when I first came across it. Win-win outcomes to social dilemmas were just what I was searching for. Here was game theory telling me that it is actually possible to achieve them, so long as we can establish genuinely stable coalitions.

One problem with establishing such coalitions is the matter of trust, as a friend's two children worked out for themselves when their doting grandparents gave one of them a bicycle for Christmas and gave the other a video game console. Unfortunately the grandparents mangled things by giving the bicycle to the one who really wanted the game machine and vice versa. It sounds like an easy enough problem to solve – just swap the gifts. But they didn't at first, because neither would go first in giving up their present to the other one, arguing, 'What if I give him my present and then he just keeps them both?'*

The children fell into the trap because they didn't trust each other sufficiently to form an alliance in which each was committed to the swap. Their parents solved the problem by threatening to take both presents away. This forced the children to form a temporary coalition, and the swap was successfully made.

The moral of this little story is that the children formed a coalition only because their parents made it worth their while to do so. In a world of selfish individuals this seems to be the main reason why we agree to form coalitions – because it is worth our

---

*They were caught in a Prisoner's Dilemma, where give-give was the cooperative, coordinated (and optimum) strategy, but keep-keep was the dominant, losing Nash equilibrium.

while to do so, or because others make it worth our while. The reward for joining may be an emotional one – the good feeling of belonging, for example, or the feeling of security that the group offers. It may also be a practical one – such as the promise of a job or a position of power, or access to resources that the person wants, or the threat of what might happen if the person doesn't join the coalition. It may even be a matter of money (promising to pay for goods, paying a sweetener to an agent, or even promising an actual bribe).

Game theorists don't make any moral judgments about such payments. They simply lump them all together as *side payments* (payments that you make to someone to keep them on your side and to stop them from leaving your coalition). Even the money that you pay to a shopkeeper is a kind of side payment to persuade them to cooperate by giving you the goods in return for the money.

Some side payments may seem just and moral. Others may seem quite unethical. Whatever the attitude, the fact is that most people will not generally cooperate to join a coalition unless they are going to get something out of it.

When more than two parties are involved, the problems multiply, but the basic principles are the same. Their choice of coalitions, though, becomes wider. Even when just three people are involved, there are three different ways in which two of them might form a coalition and gang up on a third. In larger groups, such as committees, business and social organisations, and even families, cliques inevitably form. The resultant back-stabbing, gossiping, and switching of allegiances form the stuff of many novels, and a glance through any newspaper soon reveals a substantial proportion of stories that concern the same problems.

If ants, bees, and wasps had newspapers, there would be no such stories, because they are genetically programmed to form *grand coalitions* in which all of the individuals are involved and from which they do not have the power to escape. Our individuality-preserving alternative is to offer side payments to make it worthwhile for people to cooperate in smaller groups. Of course, even then, our troubles are just starting. Once we start cooperating we must find ways to maintain cooperation, and this can be no easy task, especially when we don't *really* trust each other.

Commitment

Is there some way of ensuring that people will remain committed to cooperation in the absence of trust? In chapter 2 I argued that the most reliable way, which should work even when the parties cannot or will not communicate, is to produce a self-enforcing agreement. This generally means casting the agreement in the form of a Nash equilibrium that neither party can independently escape from without loss, so that they are trapped into cooperation. If they are able to communicate and negotiate, Nash suggested another approach in the form of a unique negotiation method called the *Nash bargaining solution*. Here I examine the ramifications of these two approaches.

Trapped into Cooperation

The Nash equilibrium can sometimes be used to lock us into a set of coordinated, cooperative strategies, because neither participant can improve their position by adopting a different strategy. The philosopher David Hume gives a nice example in a story about two oarsmen who are sitting side by side in a row-boat, each wielding one oar. As he put it, 'the two men pull the oars . . . by common convention for common interest, without any promise or contract'. They are driven to form a coalition by mutual self-interest, and the coalition is stable because it would not pay either of them to rest while the other is rowing, since the boat would then go around in circles. They are caught in a Nash equilibrium, but in this case it happens to be the coordinated, cooperative solution.

The Nash equilibrium is not always the bad guy when it comes to cooperation. As I noted in chapter 3, it often traps us in social dilemmas, but there are some situations (such as the one above, or the case of two people approaching each other on a pavement) where the cooperative, coordinated solution turns out to be a Nash equilibrium. In these situations there is no social dilemma involved – all that matters is discovering the appropriate strategies.

The ideal outcome in such cases is to discover strategies for *minimally effective cooperation* – that is, getting the job done but not putting any more energy in than is needed. In game theory terms, minimally effective cooperation is an *efficient* choice of strategies, because there is no way to rearrange things so that one or more people are better off without making anyone else worse off. (This is called a *Pareto optimal* position in economics.)

Minimally effective cooperation is the best outcome that we can hope for in many situations, whether we are brokering an international peace agreement, trying to get competitors to join in a business deal, or even just doing the housework. My wife and I had what a friend would call 'an exchange of views' on the latter subject while I was writing this chapter. The bone of contention was the amount of work that needed to be done on the house before friends came to stay with us. She thought that a lot needed to be done, while I thought that a quick vacuuming and new towels in the bathroom were all that was required before I sat down to watch a tennis match on television. But this was like a red rag to a bull, and she kept coming up with more jobs that just *had* to be done. Experienced couples will be able to work out the rest of the scenario for themselves.

This scenario was in sharp contrast to one just a few weeks later, when we adopted a strategy in which she drew up a list of jobs that she wanted me to do before a party, and I agreed to do them with the understanding that she wouldn't then expect me to do any more. A feeling of calm reigned over the house because we had negotiated ourselves into a state of minimally effective cooperation, in which we were caught in a self-created Nash equilibrium. It worked well; the state of the house was to my wife's satisfaction, and I was still able to watch some of the tennis.

Cooperative Nash equilibria don't always make things so easy, however. In many situations there is more than one cooperative equilibrium and no obvious way to choose between them. Take, for example, the pavement situation. The parties have to coordinate their movements or they might end up face-to-face again after they have stepped to the side. They could stand and discuss the issue and reach an agreement by means of negotiation, but this would be a rather exaggerated way of doing things. Most people simply watch to see what the other person is most likely to do and then move accordingly.

As I mentioned in chapter 3, a Nash equilibrium that we reach by means of such hints is called a *Schelling point*. I described an experiment that involved giving false clues to see what would happen in the absence of such a Schelling point.

The results illustrated that choosing between two cooperative Nash equilibria can be quite tricky in the absence of such a clue. What, then, of a situation where there are many such equilibria – even an infinity? Is there one Nash equilibrium that is better than the rest, that rational bargainers could reach just by negotiation?

The Nash Bargaining Solution

Such a situation could have arisen between my brother and me when we divided our fireworks. Instead of insisting on I Cut and You Choose, my father could have told us to negotiate the division between ourselves and then tell him what percentage of the fireworks we were each claiming. He might also have added the proviso that if the two claims added up to more than 100 per cent, neither of us would get any fireworks at all.

If he had, he would have been replicating the approach to bargaining that John Nash had just worked out in Princeton, twelve thousand miles away. Nash realised that if the two claims added up to exactly 100 per cent, then *any* division (apart from 100:0) leads to a situation that he would later analyse in detail as the Nash equilibrium. If, for example, I had said to my brother, 'No matter what you do, I'm going to claim 70 per cent,' and he had really believed me, then the best that he could have done would be to claim 30 per cent, and we would both have lost everything if either of us had subsequently tried to claim more.

While we were still negotiating, though, he might have responded to my claim by saying, 'Well, *I'm* going to claim 70 per cent, so stick *that* in your pipe and smoke it.' If I had really believed *him*, my best option would have been to back down and claim only 30 per cent.

Is there any rational way to resolve such an impasse? Nash proved that there is. It is the Nash bargaining solution, and it applies to any situation in which two or more parties have to negotiate to divide up a finite resource and then submit sealed bids claiming some proportion of the resource, with the proviso that if those bids add up to more than the total value of the resource, neither party gets anything. Subject to certain conditions, rational participants should always choose the division that maximises the product of their utility functions.*

In other words, the parties should look at what they would get for each possible division, compared to what they might get if they demanded more. They should then choose the division that gives the greatest yield when their two gains are multiplied together. If, for example, the total reward on an offer is £100, and the participants are only interested in the money (so that utility = cash benefit), they should rationally take £50 each, because 50 × 50 = 2,500, and for any other split the product is less (e.g., 99:1 gives a product of 99, and even 51:49 gives only 2,499).

If this sounds very far from real life, it isn't. Negotiations for the purchase of advertising media time, for example, have been found to produce Nash bargaining solutions, as have many other marketing negotiations. Nash's clever approach to rational bargaining and sharing has been used to help design a new type of

---

*Nash listed four conditions under which his bargaining solution would provide an optimal outcome:

1. That the answer divides the resource in such a way that there is none left over.
2. That the solution does not depend on how the participants assess the utility of the reward to themselves.
3. If the outcomes that they would not have chosen become infeasible, nothing else changes.
4. That the solution is unchanged if the participants change places.

auction that is used for the allocation of broadcasting frequencies. The first auction was held in the ballroom of the Omni Shoreham Hotel in Washington, D.C., in 1994, and it yielded nearly $617 million. Another held later in the year raised $7 billion, causing William Safire of the *New York Times* to label it as 'the greatest auction ever'. The continuing series of spectrum auctions (that is, auctions of frequencies from the broadcasting spectrum) is now conducted over the Internet and has yielded more than $100 billion to date.

One great advantage of this approach is that it has made tactical bidding a losing proposition. (Tactical bidding is the strategy of bidding on some frequencies to stop competitors from getting them, even though the bidder doesn't actually want those frequencies.) All participants in the original auction declared themselves highly satisfied with the outcome, in contrast to participants in Australia and New Zealand, where similar auctions conducted at around the same time, but without invoking Nash's approach, proved to be huge and costly disasters. Nash's approach is now universally acknowledged as the one that works.

The auction design has now been widely copied to sell goods and services that include electric power, timber, and even pollution reduction contracts. Its success, though, doesn't mean that game theory has all the answers. Some sceptics have even argued that it can be used to rationalise anything. Strategic analyst Richard Rumelt has argued, for example, that 'the trouble with game theory is that it can explain anything. If a bank president was standing in the street and lighting his pants on fire, some game theorist would explain it as rational'.

Management analyst Steven Postrel decided to find out whether Rumelt's Flaming Trousers Conjecture was true – and found that he could construct a perfectly reasonable, game theory-based rationale for bank presidents to publicly set their pants on fire (as a publicity stunt to attract and retain customers)! He went on to argue, however, that 'this criticism is without force. Game theory is a toolbox for constructing useful models, rather than an empirically substantive theory; its power comes from imposing logical discipline on the stories we tell'. In other words, the science is not a tool for controlling the world so much as a tool for helping us to understand it in a new and informative way. It is a guide to decision-making that gives us pointers to what is really going on, not an auto-decision maker into which we just feed the facts.

Are We Rational?

The Nash bargaining solution, for example, demonstrates that it is possible in principle to reach a fair outcome without having a sense of fairness, just by pursuing our own self-interest during negotiations in a truly rational manner to reach a uniquely best solution for all concerned. But are we really that rational? People's behaviour in the remarkably simple Ultimatum Game suggests not.

The game has been played primarily in psychological laboratories, although it has many uncomfortable parallels in real life. An experimenter gives an amount of money or other goods to someone who is then required to offer a proportion to a second person. The second person can then either accept or reject the offer. If they accept it, the money or goods are shared accordingly. If they reject it, neither of them gets anything. That's it. There is no further bargaining; it's a one-off.

What should the proposer do? His or her obvious course is to offer as little as possible, because the receiver has to accept it or get nothing. This sort of take-it-or-leave-it negotiating tactic has been widely used by the powerful to take advantage of the weak and helpless, most notably in the payment of sweatshop wages. It is powerfully represented in the 1976 film *The Front*, in which Zero Mostel plays an actor who, blacklisted during the McCarthy era, commits suicide after having performed for a pittance only to have his fee further reduced, after he has given his performance, by a cynical nightclub owner, with the words, 'Take it or leave it. No one else is going to give you a job.'

Take-it-or-leave-it is a weapon for those in positions of power. When researchers handed that power to volunteers in the Ultimatum Game, though, they received a surprise that set them right back. They found that most proposers did not try to keep as much as possible for themselves but offered around half of the total, even when real money was involved. Even more surprisingly, when receivers were offered less than 30 per cent, they often exerted their own power by rejecting the offer, even though this meant that they lost out along with the proposer. Receivers seemed very willing to cut off their nose to spite the other person's face – and not only in the affluent United States but also in countries such as Indonesia, where the sum to be divided was $100 and where offers of $30 or less were frequently rejected, even though this was equivalent to two weeks' wages!

This was totally unexpected behaviour from the cold rationalist's point of view. What was going on? One clue has come from scientific studies conducted at Princeton University and the University of Pittsburgh. Researchers used functional magnetic resonance imaging to watch what was going on in the brains of the participants when they were accepting or rejecting offers.

They found that a region of the brain known as the 'bilateral anterior insula', which becomes very active during experiences of negative emotions such as anger and disgust, also becomes active when a low offer is received during the Ultimatum Game. By contrast, a brain area called the 'dorsolateral prefrontal cortex', or DLPFC, which is known to be involved in cognitive decision-making, became very active when a high offer was made.

Game theorist Martin Nowak sees the behaviour of people playing the Ultimatum Game in terms of irrationality, saying that the game is 'catching up with the Prisoner's Dilemma as a prime show-piece of apparently irrational behaviour'. Interviews with people who have rejected low offers reveal, though, that they have done it for a reason – to punish the one who has made the low offer. The researchers who studied brain activity during the game deduced that 'the areas of anterior insula and DLPFC represent the twin demands of the Ultimatum Game task, the emotional goal of resisting unfairness and the cognitive goal of accumulating money', and sagely concluded that 'models of decision-making cannot afford to ignore emotion as a vital and dynamic component of our decisions and choices in the real world'.

So emotions need to be factored into the equation. Researchers have found that raising the stakes in the Ultimatum Game generally produces offers that are *closer* to a 50:50 split, which is hardly what one would expect if the players' motivations were based on concrete rewards alone. Maybe the sense of fairness has something to do with it. There is some evidence, though, that the sense of *fear* is at least equally important – fear that an offer will be rejected if it is too low. It is a fear that is frequently justified by reality.

These experiments show that our feelings have to be factored in as part of the gains and losses that we are trying to balance. But measuring those feelings is another matter. I would love to have watched people playing the Ultimatum Game with their heads stuck inside a giant magnet, but even such advanced scientific tools don't let us quantify feelings in the way that we can quantify money or material goods. The pleasure of punishing someone for their meanness is more easily quantified by seeing how much they are willing to give up in the way of material goods for the sake of having that pleasure, and it seems that this amount can be quite high.

When that pleasure is factored in, along with other emotional rewards or deficits, it seems that the Nash equilibrium can indeed lock us into cooperative solutions to problems in a limited range of circumstances. Reliance on external authority to enforce fair play can also help in some cases, as it did for the two children swapping their gifts. To make real progress in cooperation, though, and to avoid the seven deadly dilemmas, we need to develop more effective trust mechanisms. Only then will we be able to adopt coordinated strategies to solve problems, secure in the belief that the other party or parties will stick to the bargain and not try to do better for themselves by independently changing their strategy. To make this work, though, we need to find some way of implementing the third possible approach to commitment, which is to find genuine, compelling reasons to trust others, and to develop concrete strategies to prove to them that they can *truly* trust us. In the next chapter I review and try some out as I continue my search for strategies for cooperation.

6

Trust

ONE OF MY FAVOURITE *Peanuts* cartoons shows Linus clutching his ever-present security blanket until Charlie Brown's little sister, Sally, crawls up and distracts him with a kiss, while Snoopy grabs the blanket and runs off with it. 'If you can't trust dogs and little babies,' he sighs, 'who *can* you trust?'

Not many people, it would seem. The social dilemmas of game theory and the real world have their devastating effects because we can't, or just won't, trust each other. If we could, then many dilemmas would simply disappear. With genuine trust, we could negotiate to coordinate our strategies and produce co-operative solutions, secure in the knowledge that we could trust each other not to break agreements for individual advantage. Instead, we often act on our belief that other parties are likely to cheat, and the strategies that we work out on that basis constantly draw us into Nash equilibria.

When Sir Walter Raleigh reputedly took off his cloak and spread it across a muddy gutter so that Queen Elizabeth wouldn't get her feet wet while crossing, both of them won out through trust. He trusted that she would accept the gesture; she trusted that he wasn't playing some trick, such as pulling the cloak away at the last minute. It wouldn't work today.

I know. I've tried it. I went out into a London street on a rainy day and ceremoniously laid my jacket (an old one) over a puddle that a woman was trying to cross. She viewed my outstretched jacket with the utmost suspicion and then took a long detour to get around me and the puddle. When I repeated the experiment with other women and other puddles, the same thing happened. Not one of them would step on it, fearing some trick. Several of them even looked around for the hidden television cameras. Unlike Queen Elizabeth, they didn't trust my good intentions at all. When a friend of mine tried a similar experiment in New York at my behest, he fared even worse. Some women laughed at him, and a mistrustful policeman even asked him to move on and stop bothering people.

What could we have done to persuade them that we were trustworthy? Maybe we should have taken lessons from Lucy van Pelt, who invariably persuaded Charlie Brown that she was not going to pull the football away when he ran up to kick it. 'Look at the innocence in my eyes,' she said on one occasion. 'Don't I have a face you can trust?' 'She's right,' muses Charlie. 'If a girl has innocent-looking eyes, you simply have to trust her' – and he lands flat on his back yet again. 'What you have learned today, Charlie Brown,' she says, looking down at him, 'will be of immeasurable value to you for many years to come.'

What most of us seem to have learned is that mistrust, rather than trust, is the strategy that more often pays dividends. Sometimes we are right. More often than we realise, though, we've got it terribly wrong. We need trust. Without it, our societies couldn't function at all.

According to Barbara Misztal, author of *Trust in Modern Societies*, trust performs three functions: it makes social life more predictable, it creates a sense of community, and it makes it easier for people to work together. The trust that we offer freely to friends, family, and loved ones eases our paths through life. The communities that we live in are built on trust and often collapse when that trust goes missing. We are even happy to put our trust in little bits of paper with green printing on them. We can't eat them, build with them, ride on them, or even use them as hats or umbrellas to protect us from the elements. We nevertheless trust that complete strangers will accept them in exchange for things that we can genuinely use, like food, housing, transportation, and consumer goods. The more that we can trust, the easier and more fruitful our life becomes. Game theory tells us why, in three steps:

1. Non-cooperative solutions to problems (those that arise when we pursue our own individual interests, only to walk straight into one of the seven deadly dilemmas) occur when the participants cannot trust each other, and so cannot make credible commitments to cooperative strategies.

2. If people can arrive at a cooperative solution, any non-constant sum game (including most of our social interactions) can, in principle, be converted to a win-win game.

3. *Conclusion*: If we could find ways to trust each other, we could then find win-win solutions to many of our most serious problems.

The Origins of Trust

There is good evidence from psychological and sociological studies that most of us have an innate urge to trust. According to the pioneering developmental psychologist Erik Erikson, we face a crisis in the first year of our lives that determines how strong that urge will be. The behaviour of our major caregiver (usually our mother) determines the outcome of the crisis. If our caregiver responds predictably, reliably, and lovingly toward us, we develop a firm sense of trust. If not, we are more likely to develop a mistrust that continues through life.

The levels of trust that we experience in various circumstances depend, in part, on the hormone oxytocin, which our brains manufacture. Oxytocin is best known for its role in labour and lactation, but it also facilitates approach behaviour in many mammals, enabling them to overcome their natural avoidance of proximity to others. It is involved in pair bonding, maternal care, sexual behaviour, and the ability to form normal social attachments among many animals. Some physiologists have labelled it the 'lust and trust' hormone in these animals. Neuroeconomist Paul Zak from Claremont Graduate University conceived the idea that it might fulfil a similar role in humans of all ages, and with his colleagues he designed and performed a beautifully simple experiment to test this idea.

Their plan was to change the concentration of oxytocin in the brain and then to measure the effect of the change on a person's willingness to trust. Their way of changing the concentration was to spray some oxytocin up the subject's nose, where it could pass through the mucous membranes to enter the blood-stream and subsequently cross the blood-brain barrier to enter the brain. They compared the effect of the spray with that of a spray that contained no oxytocin.

The experimenters measured the effect of oxytocin on trust by having their subjects play a trust game. The volunteers were given a certain amount of money and told that they could either keep it or offer it to a second person. They were told that if they did make the offer, the amount given to the second person would be tripled, and the second person would be asked to give as much back to the original subject as they felt inclined to give.

If the subject trusted the recipient to be fair and give back half the final total, both would gain, but in the absence of such trust, the obvious course was for the subjects to hold on to the original amount. Those subjects who received the oxytocin became much more willing to hand over the money. The experiment showed that this was not just because the subjects were more willing to take risks but that 'oxytocin specifically affects an individual's willingness to accept social risks arising through interpersonal interactions'. In other words, they became much more trusting.

It wasn't long before an advertisement appeared on the Internet: 'Want trust in a bottle? Get Liquid Trust, the World's First Oxytocin Spray, for Proven Results!' This ethically dubious product (with which the original discoverers were *not* involved) was advertised as being 'specially designed to give a boost to the dating and relationship area of your life', as well as useful for salespeople and office managers. I wonder what a date's feelings would be if someone tried to spray oxytocin up their nose on a first encounter? I suspect that trust would be fairly low on the list, even after the spraying. As for salespeople and office managers, they would probably find themselves in court.

You can't get trust in a bottle. It comes ultimately from the overall functioning of our brains, which some scientists argue have evolved in two parallel ways – selfish and social. Oxytocin is just one of the factors that affects the complex balance between the two. On the selfish side we have the Machiavellian intelligence* that allows us to compete for mates, income, status, and more. The other side has been tailored by evolution to cope with group living, and it has been adapted to be cooperative. Opinions differ as to which of the two has driven the huge increase in brain size that our species has undergone over thousands of years. One thing is clear, however – the Machiavellian side of our brains, where we act out of pure self-interest regardless of the interests of others, is the one that leads us into social dilemmas, while the cooperative, social side provides us with ways to escape from them.

The Evolution of Trust

Indeed, the social side of our brains is fuelled by trust. When we look at how trust works, though, it is very hard to see how our ability and wish to trust could have evolved in the past, let alone how we could encourage it to evolve in the future. Evolution strongly favours strategies that minimise the risk of loss, rather than those which maximise the chance of gain. Trust, however,

---

Machiavellian has become a synonym for all that is devious, underhanded, and reprehensible, but Machiavelli's primary message to those who want to win and maintain power was that 'it is far better to earn the confidence of the people than to rely on [force]'. Trust, to Machiavelli, was the central issue, even if the methods that he suggested for winning it were sometimes based more on practicality than on morality.

does just the opposite. If we offer trust, we are taking a risk that the trust may be betrayed. If the risk pays off, we may gain a lot, but if it doesn't, we can lose out in a major way. Betrayal of trust can lead to the failure of a relationship, the loss of money, or even the loss of health or life if the trust that we put in a particular treatment or medication turns out to have been misplaced. In the life of a species, misplaced trust can even be a factor in extinction, as it was in the case of the dodo, which would let people walk right up to it to hit it on the head.

Game theorists call the offering of trust a *pay off-dominant strategy* – that is, a strategy in which the user aims for the maximum possible pay off from a given situation. It is the sort of strategy that our cat, Yasmin, follows at meal times when she refuses a dish of lamb or beef and sits looking wistfully upwards, hoping that we might relent and give her a plateful of tuna instead, or even a dish of pheasant or guinea fowl more suitable to her regal name and status. The three cats next door, on the other hand, follow a mistrustful *risk-dominant* strategy (one in which the avoidance of risk is the primary objective) by gobbling up all of their food as soon as it is put on their plates, rather than running the risk that the other cats might steal it.

Over time, those members of a species that use risk-dominant strategies will tend to prosper, while those that use pay off-dominant strategies are unlikely to survive. If our cat went next door for her meals, for example, she wouldn't last long unless she changed her approach.

Mistrust is risk-dominant, while trust is pay off-dominant. This means, in simple evolutionary terms, that mistrust should always predominate. Natural selection has seen to it that those with the most highly developed sense of mistrust are those that have the best chance to survive and pass that sense of mistrust on to their progeny. Mistrust for these animals is the evolution-arily stable strategy.

There are some circumstances, though, in which trust con-veys an evolutionary advantage. It is important, for example, within small social groups such as families and tribes. Evolution thus lands us with contradictory urges to trust and to mistrust. The two urges fight a perpetual tug-of-war in our brains, with learning and life experience as the referee. Cooperation can only happen when there is trust, but there is a second condition – the trust must also turn out to be justified. To learn how to cooper-ate, then, we must not only learn how and when to offer our trust to others; but also learn how to win it from them.

Learning how and when to offer trust is tricky enough in it-self, because it is not always easy to distinguish between genu-ine commitment and empty promises. Some people claim that they can tell the difference through reading a person's body language, but experiments have shown that such beliefs are usually without foundation. In one experiment, British psy-chologist Richard Wiseman arranged for a well-known TV pre-senter to record two TV clips. In one, the presenter described his favourite film truthfully, and in the other he lied, claiming another film as his favourite. Wiseman then asked viewers to pick which interview showcased the man telling the truth and which they thought was the lie. The result? Only half of the people who viewed the clip got the answer right – no better than a statistical guess.

That's a pretty worrying result for those of us who rely on intuition to distinguish truth from lies. Intuition can let us down badly, as shown by the fact that so many of us still fall for confidence tricks of one form or another. Here are some of my favourites:

The Barred Winner: A man approaches you outside a casino with a bag of high-value chips, saying that he has been thrown out and can't cash them but offering a percentage if you will do it. He demands some security, though, such as your wallet. When you go into the casino, you find that the chips are fake.

The Hidden Money Internet Scam: This is just one of many hidden money scams, in which you are made to think that you will gain money by helping someone to retrieve huge sums that they can't access themselves.

The Romance Scam: In another Internet favourite, a lonely person is led to believe that they have found true love, after which the 'lover' asks for money to help pay fictitious debts or travel to join their victim, which of course they never do. (If they did, the victim might be in for a surprise, because men often pose as women in this scam.)

Get-Rich-Quick Schemes: These include chain letters, pyramid schemes, fake franchises, wealth-building plans, advice from unqualified self-help gurus, and investment in useless products – the list is endless.

Of course, the oldest confidence trick of all was invented by one William Thompson in 1849. Thompson, dressed in genteel fashion, would approach wealthy New Yorkers and after a brief conversation, during which he would bemoan people's lack of trust in him, he would ask, 'Have you confidence in me to trust me with your watch (wallet, etc.) until tomorrow?' The victim, placing confidence in Thompson's honesty, would lend him his belongings, only to have Thompson never return.

It's hard to believe that anyone would fall for this, but apparently many people did, misled by Thompson's appearance to trust their intuitive judgment that he was honest. I inadvertently pulled off a similar confidence trick when I was working for a national research organisation in Australia and walked into the library of a branch where I was not known, and where I was not carrying any identification. Even so, the librarian let me leave with several valuable books. As I walked out, I heard someone ask her, 'Who was that?' 'I don't know,' she said, 'but he seemed so confident.'

Her intuitive judgment worked in this case – I eventually returned the books. But intuitive judgment is often insufficient, or just plain misleading, when it comes to knowing whom to trust. Is there some way that we can do better?

Credible Commitment

I found that the game theorist's answer to the question of trust is to use *credible commitment* as a touchstone, which involves each party demonstrating its commitment in a way that gives the others good reason to believe in it, even if they do not trust the party itself. Lucy, for example, might have offered to have one hand tied behind her back so that she couldn't physically pull the ball away as Charlie Brown ran up to kick it. Charlie Brown would then have had some grounds, other than the look in her eyes, for believing her.

Game theorists offer two basic ways for you to demonstrate credible commitment without the necessity for underlying trust. Both involve limiting your own options *in a way that the other party knows about*. Lucy would have been limiting her options, for example, by allowing one hand to be tied behind her back. The point of doing it, though, would have been to make Charlie Brown confident that she would not pull the football away – not because she wouldn't, but because she couldn't.

The two basic ways are:

1. Make it too costly for you to change your mind later.

2. Go even further and deliberately cut off your escape routes, so that you have no chance of backing out.

Making It Too Costly for You to Change Your Mind Later

There are six broad strategies that we can use, with outcomes that can vary from the hilarious to the horrendous if we happen to change our minds and fail to deliver on a promise or a threat:

1. ***Put yourself in a position where your reputation will be damaged if you do not deliver:*** We do this much more often than we realise. When actors undertake a stage role, for example, they are implicitly putting themselves in a position of being unlikely to be offered any more roles if they don't turn up for each performance of their current role. Threats to punish one's children, or offers to reward them, also come into this category. My parents took my pet dog Rusty away when I was a child because he was digging

up the garden. I didn't make an inordinate fuss because I believed their promise that they would give me chickens instead. I never got those chickens, and I never believed their promises thereafter.

2. **Move in steps:** Breaking a promise or threat into a series of steps means that when you get towards the end, most of the promise or threat will have been fulfilled, as happens when homeowners or developers pay builders at the end of each completed phase of a project. But there is a trap here. If you *know* that it is the last step, you may be tempted to renege. A developer, with the project completed, may refuse the last payment, leaving the builder short, or with the stress and cost of taking the developer to court. A tenant may skip without paying the last month's rent, as has happened to me as a landlord more than once. The message is clear: make the steps (or at least the last few steps) as small as possible so as to minimise the risk of loss.

3. **Work in a team or a group:** This is another way of putting your reputation on the line, because letting others in the group down can do you future damage when they then fail to trust or accept you. You could even be left out of the team entirely, as happened to me when I played lazily and without commitment on my church soccer team. I felt that there was no worse punishment! (Roman soldiers might have disagreed with me, since death was the punishment for anyone hanging back in an attack. To make this draconian punishment work, failure to kill someone who hung back was also regarded as a capital crime!)

4. *Cultivate a reputation for unpredictability:* This sounds crazy, but it can work in unexpected ways. If you can be trusted not to be predictable, you can sometimes benefit. When I was a science undergraduate, a fellow chemistry student who had a lucrative scholarship from a leading paint company turned out to be a real nutcase, on one occasion pouring ether down a sink at one end of a lab and then holding lighted matches to a sink at the other end to see how long the vapour would take to get through and explode. The company that was paying for his education heard about it, and his reputation for unpredictable behaviour in the lab meant that they released him from his agreement to work for them for relatively low wages when he had finished his degree.

5. *Enter into a contract:* Some contracts are binding, as Faust discovered when he entered into a contract with the devil. Most contracts, however, can be subject to renegotiation. To make them stick, they often need something extra, such as a penalty clause. The person or body who enforces the clause must also have a good reason to stick to their responsibility. Penalty clauses are of little use if, for example, a local planning officer can be bribed into approving a shoddy piece of building work, for example, even though that work does not meet the standards of the contract.

6. *Use brinkmanship:* 'I'll shoot unless you pass over that bag full of money!' screams a man standing at the counter of a bank. How realistic is his threat? It doesn't really matter, because the outcome will be so drastic if he carries it out.

That's the essence of *brinkmanship*, a term coined by U.S. presidential candidate Adlai Stevenson at the height of the Cold War in 1956. Stevenson used it to criticise Secretary of State John Foster Dulles for 'bringing us to the edge of the nuclear abyss'. I mention it here only to complete this list of ways to demonstrate credible commitment by making the cost of escape too high. It is certainly the least likely of the lot to lead to genuine cooperation!

Deliberately Cutting Off Your Escape Routes

There are three broad ways of doing this, each scarier than the last when it comes to limiting your own options:

1. *Use a mandated negotiating agent:* With a legally binding contract, that agent is the law. But there are many agreements we enter into that are not legal contracts but that are contracts nonetheless. When my brother and I divided up the household jobs between us, our verbal agreement was a contract, and it was enforced because we had a mandated negotiating agent – our father!

2. *Burn your bridges:* We do it whenever we post a letter, press the send button for an e-mail, turn off our mobile phones after leaving a message, and even after we write our wills. Once we've done it, that's it. We've made a commitment, and that commitment is credible because there is no going back.

   There are many ways to burn your bridges. The philosopher Ludwig Wittgenstein found an unusual approach when he decided that he wanted to live an ascetic life, unencumbered by the burden of money. He deliberately divided his considerable fortune among his relatives in such a way that they couldn't give any of it back to him without being severely penalised for attempting to do so. When Hernán Cortés scuttled his ships, he limited his option to sail away from Mexico in dramatic fashion.

Two friends of mine found another way when they decided to skydive. Both of them got an attack of nerves, with each saying to the other, 'If you go first, I'll follow you.' Neither would really trust the other to follow until they hit on the idea of offering credible commitment by each taking a grip on the other's wrist, so that when one jumped, the other was forced to follow.

3. *Put your decision in the hands of fate:* This does not mean tossing a coin or rolling the dice as much as it does taking some action and awaiting an outcome that is both uncertain and irrevocable. The game of Russian Roulette in Ingmar Bergman's 'exquisite carnal comedy', *Smiles of a Summer Night*, provides a classic example. Two men, competing for the love of a woman, decide to settle the matter by playing this dangerous game, taking turns to fire from a gun in which one chamber is loaded. The audience can only see the outside of the summer house in which the challenge is taking place, and from which (after an unconscionable interval) there comes a loud bang. One of the men then comes out, laughing uproariously – closely followed by the other, his face covered in black powder. The first man had loaded the gun with a blank.

The film *Dr. Strangelove* provides a particularly powerful example of what can happen when you limit your own options by leaving an outcome to chance but failing to let others know about it. The Soviets' doomsday machine was a way for them to limit their own options in the case of war. Unfortunately, they have not had time to communicate its existence to the Western powers before the delusional Brigadier General Jack D. Ripper takes it into his head to launch a first-strike nuclear attack. Result: catastrophe, as the gloriously named Major 'King' Kong (played by the equally gloriously named actor Slim Pickens) rides a nuclear bomb earthward, its phallic positioning between his legs representing exactly what he was about to do to the world.

Generosity and Altruism

Credible commitment works, even in the absence of underlying trust between parties. If we could have such underlying trust, though, the problems of cooperation would often be much easier to resolve. How can we go about it?

One way in which we can gain trust is by showing altruism and generosity toward others without the expectation of reward. Generosity is often considered to be a subset of altruism. Most of us understand altruism to mean helping another at a cost to oneself, while generosity, in addition, implies 'liberality in giving'. The famous Scottish music hall singer Harry Lauder was not in favor of either. He was once confronted by a charity collector in an Edinburgh street with the peremptory demand to 'give till it hurts'. 'Madam,' he is said to have replied, with tears in his eyes, 'the verra idea hurts.'

It doesn't hurt most of us, though, because altruism, and even generosity, bring their own rewards. This sentiment was encapsulated in a sign I recently saw on a Sydney bus that read, 'Consider others. Feel good about yourself.' That feeling may be related, like trust, to the concentration of oxytocin in the brain. It would be absurd reductionism to say that brain chemistry and physiology alone account for our feelings, but they obviously play a substantial part. There is strong evidence, for example, that charitable giving activates reward regions in the brain. Charities recognise this response, and they (quite reasonably) capitalise on it.

The good feeling of making a contribution certainly motivates many scientists, and many scientists make financial and other sacrifices just to be involved in the scientific process. Our reward lies in communicating with and learning from other scientists, but there are other benefits, whose importance varies from person to person. One benefit is the pleasure of understanding, which drives most of us. A second is the acknowledgment of peers. For some, there is also the financial reward that can (occasionally) come from a successful discovery or invention. The best reward of all for many of us, though, is the altruistic feeling of having made a contribution.

Leaving our footprint in the sand means communicating our discoveries freely. Sharing our data and ideas creates a strong atmosphere of trust among scientists, which is why it is so shocking when scientists cheat in the hope of making a deeper mark. One scientist did this literally, setting the field of transplantation back by years in the process, when he claimed to have been able to transplant patches of skin from one (black) mouse to another (white) mouse. All he had in fact done was use an indelible marker to draw black patches on the white mouse.

The Trust Bond

Frauds are usually discovered and exposed because of the openness of science, which allows claims to be questioned and checked. Trust is maintained because scientists form a cohesive social group that is held together by trust, which plays a similarly important role in many cultures. In Japan, for example, according to Francis Fukuyama in *Trust*: 'Networks based on reciprocal moral obligation have ramified throughout the Japanese economy because the degree of generalized trust possible among unrelated people is extraordinarily high. . . Something in Japanese culture makes it very easy for one person to incur a reciprocal obligation to another and to maintain this obligation over extended periods of time.'

The same applied to the early days of Australian settlement, only it was referred to as *mateship*, defined as 'a code of conduct among men stressing equality and fellowship'. It was a survival mechanism for a harsh environment, maintained because mates did not let each other down, no matter what the circumstances. What held it together was not the fact that people were willing to offer trust; it was the fact that people were willing to earn it through putting others before themselves.

The flip side of the mateship coin was (and still is) the chauvinism, jingoism, and racism that could emerge from suspicion of outsiders who haven't earned that trust. The human tendency to mistrust outsiders has been the subject of study by political scientist Robert Putnam, author of *Bowling Alone*. Putnam produced hard evidence that mistrust increases with social diversity in communities. This is particularly disappointing for those of us who believe that diversity of culture can foster understanding and trust, promote creativity, and boost economic productivity in the long run.

The shock from Putnam's research was that out-group suspicion did not produce more in-group cohesion. Quite the opposite. When people from widely different communities were asked how much they trusted each other, he found that not only did people of different ethnicities trust each other less but also the level of trust between people of the same ethnicity fell off as well with increasing ethnic diversity in the overall community.

In a 2006 lecture Putnam argued that we need to learn how to become more comfortable with diversity:

> Ethnic diversity will increase substantially in virtually all modern societies over the next several decades, in part because of immigration. Increased immigration and diversity are not only inevitable, but over the long run they are also desirable. Ethnic diversity is, on balance, an important social asset, as the history of my own country demonstrates. In the short to medium run, however, immigration and ethnic diversity challenge social solidarity and inhibit social capital. . . Immigrant societies have overcome such fragmentation [in the past] by creating new, cross-cutting forms of social solidarity and more encompassing identities.

That's one approach to developing trust within communities, and sometimes it seems to work against all the odds. When my wife and I visited Croatia in 2007, we passed through villages where every house was pockmarked with bullet holes as a result of the long-running Serb–Croat conflict of the previous decade. (Just imagine living under those circumstances in your own town.) Amazingly, those villages have now been reinhabited by the mixed communities of Serbs and Croats who used to live there. Seeing themselves as members of the same village community has, in the long run, acted as a social adhesive that has resisted the dissolving power of ethnic bitterness and suspicion. It seems that Putnam's 'cross-cutting [form] of social solidarity and more encompassing identit[y]' really did work in this instance as a tool to promote trust and cooperation.

The Mistrust Barrier

'More encompassing identities' in the wider world, though, seem to have a significant downside, because identification with a group invariably seems to produce suspicion, mistrust, and looking down on those who do not belong to the group. This is reflected historically in the fact that ethnic, cultural, and religious differences have been major sources of conflict.

Influential thinkers in the first half of the twentieth century argued that world government was the only way to avoid such conflicts. Game theorists would call this a 'grand coalition' of all countries. Whatever you call it, it is surely impractical. The notion of every country, ethnicity, and religious creed pulling in the same direction is the stuff of fiction. Game theory tells us that different groups often believe that they can do better by cheating on cooperation to pursue their own goals and land up in the Prisoner's Dilemma and other social dilemmas as a result. There is also just too much mistrust, which bedevils institutions like the European Parliament and the United Nations, often rendering them impotent. The European Union's Charter of Fundamental Rights contains a 'solemn proclamation' of common values and human rights that has little legal force, because individual nations are unwilling to trust others with such force. The United Nations Charter professes a determination to 'save succeeding generations from the scourge of war', 'reaffirm faith in fundamental human rights", establish "justice and respect' for international law, and 'promote social progress and better standards of life'; but the institution fails much more often than it succeeds, judging by the prevalence of war and the abuses of human rights in so many countries that theoretically subscribe to its charter.

Many factors come into play when we are talking about trust – education, moral leadership, recognition of the rights of others, and overcoming our inbuilt psychological barriers to acceptance are just a few. Where game theory comes into the picture is in constructing and honing strategies that can lead to trust. Apart from those I listed earlier, there are two others – the use of ritual and the *offer* of trust itself. Both fulfil the game theorist's requirement that they lead to credible commitment.

Ritual

One strategy for obtaining trust is to publicly limit your freedom of action by turning the limitation into a ritual. Rituals can be very powerful, especially when social pressures or religious beliefs are involved. Naturalist David Attenborough noted a particularly interesting example when he visited the Pacific island of Vanua Mbalavu early in his broadcasting career. '[We filmed] a little known ritualised fishing ceremony . . . A great number of people, swimming continuously for hour after hour, stirred the mud, releasing the gas [hydrogen sulphide] and making the waters slightly acidic . . . Almost immediately the lake was alive with fish leaping from the surface. The advantages of ritualising such an event and putting [it] under the control of a priest are plain. The lake, being comparatively small, could have been easily fished out if there was no limitation.'

This particular ritual had a very specific purpose: the conservation of the lake's supply of fish. Early anthropologists such as James George Frazer interpreted all human rituals as having such practical purposes, but others have disagreed. Wittgenstein, for example, argued that Frazer had ignored the expressive and symbolic dimensions of rituals, and claimed that they could be fully understood only by attending to the inner meaning that they already have in our lives.

The balance of current evidence suggests that our public rituals serve both purposes. They enable public emotional expression and commit the participants to specific goals. Many marriage ceremonies, for example, fulfil the emotional wish to publicly express a feeling of love and also commit the parties to certain practical obligations. In earlier times social pressures ensured that this commitment was credible, even though some of the commitments were not ones that we would now make. Few women, for example, are now likely to hand over legal title to all of their worldly goods to their husbands at the moment of marriage.

In some cases, however, old rituals retain their force. In England, for example, gypsies at Somerset's annual Priddy Fair still seal the sale of a horse with a slap of the hands, and woe betide the man who subsequently tries to back down on the deal. In other countries, once an offer to purchase a house for a specific price has been accepted by means of a ritual handshake, a binding contract is formed, and the seller is legally bound to sell the house to that purchaser for that price.

In both of these cases the commitment is *credible* – in the first, because social pressure provides sanctions against those who cheat; in the second, because legal pressure provides the sanctions. Acting in the knowledge that you will be subjected to such pressures is one way to demonstrate credible commitment. But sometimes it doesn't need pressure at all. Just showing trust can be enough.

Offering Trust

Relationship counsellors place strong emphasis on the role of trust in close relationships. It plays two roles: acceptance ('Can I trust this person to accept me?') and commitment ('Can I trust this person to honour his or her commitment?').

Trust plays the same two roles in our wider social relationships. One surprising way in which we can display credible commitment is by showing someone else that we are willing to trust them, even when that trust has not been earned. Such an action can often initiate a cycle of trust by motivating others to show trust in return. The political theorist and philosopher Philip Pettit speaks of this as 'the motivating efficacy of manifest reliance'. Philosopher Daniel Hausman calls it 'the trust mechanism'. Whatever you call it, it is increasingly regarded as a major factor in our affairs, both in the functioning of economies and in the wider context of group cooperation.

Sometimes we offer trust without realising it. We are unwittingly showing trust in our fellow human beings when we lose something in a public place and hope that the person who finds it will be honest enough to return it. *Reader's Digest* conducted an experiment to find out how justified our hopes might be. Researchers left 960 mid-priced mobile phones in busy cities around the world and then rang them from a distance while watching to see if anyone would pick them up, answer the calls, and return them to the owners. Amazingly, a total of 654 were returned, suggesting that the trust mechanism really has something going for it.

Inhabitants of the city of Ljubljana in Slovenia came out tops for trustworthiness, with twenty-nine out of thirty phones returned. New York wasn't far behind, with twenty-four returns. Disappointingly, my home city of Sydney registered only nineteen, but at least we were ahead of virtue-proclaiming Singapore, which registered only sixteen, and Hong Kong, which registered only thirteen.

The reasons that people gave for returning the phones were very revealing. The most common reason was that they had themselves once lost an item of value and didn't want others to suffer as they had. Parental issues also featured in two different ways. An almost-destitute Brazilian woman explained, 'I may not be rich, but my children will know the value of honesty', while a young Singaporean explained that 'my parents taught me that if something is not yours, don't take it'.

In terms of the game theorist's concept of utils, these explanations make sense – the good feeling of returning the phone, or the bad feeling of keeping it, outweighed the material value of the phone for these people. In terms of the practical implementation of the trust mechanism, the results also promise considerable hope. It's just a matter of choosing the circumstances so that the odds are in your favour when you offer trust.

That judgment is usually a matter of experience, but it is surprising just how often an unconditional offer of trust can evoke trust in return. A colleague who moved from academia to industry discovered this when she was sent on a one-week bonding exercise with a group of strangers from her new job. As soon as they arrived, they were all told to stand on a large log that bridged a muddy stream, and my friend (who was at the end of the log) was then told to find a way to get to the other end without falling in. She could only do it by trusting each of the others in turn to support her as she made her way across, which she said was one of the most unnerving experiences of her life. But, as many who have participated in this sort of exercise will know, it somehow worked. Game theorists might say that it worked because the offer of trust stimulated others to trust in return.

The cycle of offering trust in order to have trust returned constitutes a closed chain of reciprocal logic ('I will trust you to trust me to trust you to trust me . . .'). If conditions are right, the trust will then grow and flourish. Philip Pettit describes the process: 'Trust materializes reliably among people to the extent that they have beliefs about one another that make trust a sensible attitude to adopt. And trust reliably survives among people to the extent that those beliefs prove to be correct.' This is a circular piece of logic, not unlike Anselm's argument for Christian belief ('I do not seek to understand that I may believe, but I believe in order to understand'). Anselm chose to enter the circle by offering belief without understanding. In the circle of trust, game theory suggests that it is often best to enter the circle by offering trust without experience of whether the recipient can really be trusted.

In offering trust as a way of demonstrating credible commitment, you are playing the odds of gaining trust in return against losing out if the other party proves to be untrustworthy. Just the action of trusting can tip the balance, because it means that the other party has already gained something (your expressed good opinion of them), which they will not want to lose (in game theory terms, this is their reward). Even if they have done nothing to warrant that opinion, the fact that you have offered it swings the balance in your favour and can actually make them more trustworthy. In business, for example, trusting a person with some responsibility can actually make them more responsible.

Trust is especially important in the relationship between counsellor and client. My wife is a counsellor who uses the person-centred approach of Carl Rogers, in which unconditional positive regard for the client is all-important. On several occasions I have attended person-centred workshops with her and experienced the effects of genuine trust based on the Rogers approach. People simply sit in a circle and offer confidences if they feel so inclined. Once people see that others are willing to trust them by sharing personal experiences, they become willing to share their own confidences in return. To my intense surprise, I have on several occasions even found myself offering my own confidences, induced to trust others with them by the fact that they have trusted me.

Offering trust works in some unexpected circumstances. I belong to an organisation called BookCrossing. Members leave books that they have read in public places so that others can find and enjoy them and then pass them on to others. A message in the front of the book asks the finder to do this and gives a website address readers can go to and post their comments. Most books get passed on, and some have been through dozens of readers, eventually finding their way back to their original owners!

The efficacy of the trust mechanism can depend very much on circumstances. I don't imagine, for example, that the BookCrossing approach would work very well for the sharing of cars. It doesn't even work for bicycles without severe safeguards. In one community bicycle programme that was tried in Cambridge in 1993, bicycles were made available for people to use freely around the city and then leave for others to use. The programme didn't last long; all three hundred bicycles were stolen on the first day, and the programme was abandoned.

Many people believed that the programme failed because professional bicycle thieves are prevalent in Cambridge, and the last thing that such thieves care about is what others outside their closed circle might think of them. Similar programmes have worked well elsewhere, partly because the lessons of Cambridge have been learned, and safeguards (such as fitting bicycles with electronic identification tags) have been put in place to increase the chance of penalty and reduce the chance of reward for cheating.

It is often possible to overcome the mistrust barrier and to find strategies that will evoke and maintain trust. For the long-term evolution of cooperation, though, we still need to explore additional strategies. In 1986, game theorist Anatol Rapoport uncovered another missing piece of the puzzle, in the form of the strategy of Tit for Tat, in which the parties respond in kind to the actions of others, cooperating if others offer cooperation and retaliating with non-cooperation if others have cheated. It works well when the parties concerned come into repeated contact with one another. Cheating might pay as a one-off, but it is less likely to pay if the victim has a chance to retaliate. Many species use this tit-for-tat mechanism in one form or another as a way of maintaining trust in a group.

Tit for Tat can lead to ongoing you-scratch-my-back-and-I'll-scratch-yours cooperation, but it can also lead to the escalation of conflict in the form of 'an eye for an eye and a tooth for a tooth', as it has, for example, in many current civil and international conflicts. Making the strategy produce cooperation rather than escalation is a problem to which game theorists and others have given considerable thought. In the next chapter I report my investigation of their results and the conclusions that we can draw from them.

7

Tit for Tat

MY SOCIAL TRAINING AS A CHILD was shaped by two frightening characters from a Victorian children's story. Their names were Mrs. Doasyouwouldbedoneby and Mrs. Bedonebyasyoudid, and they appeared in Charles Kingsley's *The Water Babies*, which my parents had given to me for my seventh birthday. Their moralities were very different, but both were ultimately based on Tit for Tat – a payback strategy that comes into play when two people or groups are likely to meet repeatedly. Game theorists have found that such repeated interactions are an important key to finding cooperative solutions for the seven deadly dilemmas, because the threat of future retaliation can deter cheats, and people are more likely to cooperate with you in the future if you have cooperated with them in the past.

Mrs. D and Mrs. B epitomised these two approaches, the first offering the carrot of cooperation, the second the threat of retaliation. In *The Water Babies*, these two alarming ladies act as moral guides to a little chimney sweep called Tom, who has fallen into a river and been turned into a water baby. Mrs. D alarmed me because she was uncomfortably like my mother, always pushing the Golden Rule: 'Do unto others as you would have them do unto you.' She would never punish Tom directly when he broke that rule but was adept at emotional blackmail, simply letting Tom know how much he had upset her by his latest infraction of the rules and then leaving him to worry about his own badness. I still dream about her sometimes.

Mrs. B alarmed me for a different reason. She was a strict disciplinarian who reminded me of my nanna, a ferocious old lady who sniffed out evil and punished it with all the zeal of an Old Testament prophet. Unfortunately, the evil that she sniffed out was usually mine.

She sniffed it out literally on one occasion, when I had borrowed my father's pipe to try it out in privacy behind a hedge. She chased me three times around the garden at a time when all I wanted was solitude and repose, and it was only by dint of furious effort that I was able to climb the fence that separated us from the Presbyterian church next door, where I was flamboyantly sick in a bed of hydrangeas while she hung over the fence saying, 'Wait till you get home.' When I eventually slunk back to the house, she was there waiting, with father's pipe refilled and a box of matches in her hand. She made me smoke it right through, hoping to cure me of the dreadful habit. I often wonder if I took up pipe smoking later in life just to spite her distant memory.

Mrs. D and Mrs. B represent two extreme approaches to the problem of interacting with others. Mrs. Doasyouwouldbedoneby represents the ethic of reciprocity (otherwise known as the Golden Rule), which has been advocated by philosophers from Socrates onwards as a basis for practical morality, and which is advocated by most of the world's major religions. Jesus propounded it in the Sermon on the Mount when he said, 'Do unto others what you would have them do unto you.' The Prophet Muhammad, in his last sermon, admonished believers to 'hurt no one so that no one may hurt you'. Confucius said in *The Analects*, 'Never impose on others what you would not choose for yourself.' The Dalai Lama put it in a different, thought-provoking form when he said, 'If you want others to be happy, practise compassion. If *you* want to be happy, practise compassion.'

The ethic of reciprocity is a statement of morality in which many of us believe, regardless of whether we have religious faith or not. Many philosophers have advanced it. Pythagoras said, 'What you wish your neighbours to be to you, such be also to them', while the German philosopher Immanuel Kant made an even stronger statement when he made it an example of the *categorical imperative*: 'Act only according to that maxim whereby you can at the same time will that it should become a universal law.' The categorical imperative was, according to Kant, an absolute, unconditional requirement that exerts its authority in all circumstances, and that is both required and justified as an end in itself.

The ethic of reciprocity provides a guideline for how we would wish to behave, regardless of how others respond. In Mrs. D's hands, it was also a guideline to practical strategies. 'If you want someone to trust you,' she was effectively saying to Tom, 'the best thing is to show that you trust them first. If you want someone to love you, your best approach is to show that you love them. If you want someone to cooperate with you, try cooperating with them."

Mrs. D's strategy was based on an optimistic assessment of human nature, which was in line with Kingsley's position as a social reformer who believed in the essential goodness of people. Mrs. B took a much more sceptical view of human behaviour and human values. 'Be done by as you did' was an approach based on fear. 'You can't really trust anyone,' she said, in essence, 'so if you need cooperation, the best way to get it is to threaten punishment for those who don't cooperate. If what you are after is obedience and getting others to conform to your rules, the best way to get that is the threat of punishment as well.'

Both of these approaches have evolved in nature as means of obtaining and maintaining cooperation in circumstances involving repeated interactions between individual animals. American brown-headed cowbirds, for example, use Mrs. B's retributive tactics in a protection racket. 'Raise my chick, or your eggs get it' is their threatening message to the warblers that inhabit the swamps around the Cache River in southern Illinois. When a warbler lays its eggs, a cowbird will come along to lay its own egg right alongside. If the warbler raises all the chicks, including that of the cowbird, all is fine. If the warbler rejects the cowbird's egg, however, the cowbird will retaliate by returning to the nest to eat or otherwise destroy the eggs of the warbler.

Rats, in contrast, use Mrs. D's 'Do unto others as you would have them do unto you' – and it works. When rats in a cage pull a lever that releases food for a rat in an adjacent cage, the rat in the adjacent cage becomes much more inclined to pull a lever in its cage so as to feed another rat. Rats, in other words, are swayed by the kindness of strangers to act with kindness themselves, and the whole population of caged rats eventually becomes more altruistic.

Game theorists call the evoked behaviour 'reciprocal altruism', and rats are not the only animals to use it. Vampire bats will feed blood to others that haven't managed to find any during the night, and the bats who have been fed then remember and return the favour. Chimpanzees will offer to share meat with others, even when they are not related, and will go out of their way to help an unfamiliar human who is struggling to reach a stick, just as a small toddler will do.

The strategies of Mrs. B and Mrs. D have both contributed to the evolution of cooperation in nature, but which one should we choose for ourselves? Both of them involve an element of risk. If we use Mrs. D's Golden Rule, we run the risk that others will not join us in reciprocal altruism by doing unto us as we have done unto them. If we adopt Mrs. B's threat of punishment and retaliation, we risk an ongoing cycle of retaliation and counter-retaliation if the other party does not succumb to the threat.

That risk can be very real, especially if one of the parties is an aggrieved child with an injured sense of fairness. When my Nanna forced me to smoke the pipe, I retaliated by putting a frog in her bed. She retaliated in her turn by telling my father, with painful consequences to me. I am not ready to reveal what happened next, except to say that when I launched a rocket into her bedroom, it was not quite as accidental as I made it out to seem in chapter 2.

Cycles of retaliation have their beginnings when someone feels aggrieved. I experienced an amusing example when I was a research scientist in a government organisation. Some technical staff had been appearing late for work. The management decided that the answer was to have an attendance book for them to sign when they arrived, but some of the technical staff felt aggrieved that research scientists weren't being required to sign the book as well. To egalitarian Australian eyes, there was a clear answer – steal the book. The management responded with dire threats of sanctions if the book wasn't returned. It duly reappeared, and to make sure that it didn't disappear again, the management nailed it to a sturdy wooden table. The next day the table and the attached book were both stolen. There was no further mention of the book.

Unfortunately, in the adult world, cycles of retaliation and counter-retaliation can lead to more serious consequences, including messy divorces, ongoing sectarian violence, terrorism, and war. Suicide bombings in the Middle East are responded to by missile attacks, which are responded to by yet more bombings, in an endless cycle of violence. Nobody ever really gets the last word in such tit-for-tat cycles. If we are going to use Mrs. B's threatening strategy of punishment and retaliation to establish and maintain cooperation, we need to find some way to break such cycles or stop them from starting.

Breaking the Cycle

The obvious way to stop a cycle of retaliation and counter-retaliation is for one side to stop retaliating. 'Always forgive your enemies,' said Oscar Wilde, 'nothing annoys them so much.' It annoys them because it removes their justification for continuing with the fight. It certainly shocked and annoyed my nanna when I came home from Sunday school, full of pure and beautiful thoughts after hearing a sermon on forgiveness, and proceeded to forgive her in front of my parents for some punishment that she had inflicted but which they didn't know about until that moment. My nanna and I never really got on, but the cycle of punishment and retaliation stopped then and there.

Another way to break the cycle is with an apology, which in my own marriage takes the form of a hug and the words 'I'm sorry.' It can take some doing, as anyone in a relationship will know, but with us it is an agreed-upon strategy, to be used as soon as one or the other of us realises that we have been sucked into a cycle of recrimination and counter-recrimination.

If the cycle is not broken, recriminations can go on for a very long time, as the history of the stolen generation in Australia demonstrates. Successive governments between 1900 and 1970 pursued a policy of forcibly removing part-Aboriginal children from their families and placing them with white foster parents or in orphanages. It was all done with the best of intentions, to give the children a 'better' chance in life, but the effects on that generation of children and their families (as shown, for example, in the film *Rabbit-Proof Fence*) were profound. Successive governments have refused to apologise for this shameful episode in Australia's history, provoking a cycle of recrimination and justification, but the present government has grasped the nettle and offered an unconditional apology to the individuals and families affected. With this step a great wound has begun to heal. Perhaps other governments, and other splintered societies, could take note.

It would be best, of course, if such cycles never started. Mrs. D's ethic of reciprocity aims to nip cycles of retaliation and counter-retaliation in the bud by taking pre-emptive action. 'Don't do anything that might provoke retaliation,' Mrs. D advises, 'but act toward the other person as you would like them to act toward you in similar circumstances.' It's something that we do quite often. The fact that we do so has been dubbed the Samaritan paradox, because it involves the behaviour illustrated by the parable of the Good Samaritan (Luke 10:25–37), who behaved kindly toward a stranger even though he knew that he was unlikely ever to meet the stranger again. It is hard to see how this sort of altruistic behaviour, offered at personal inconvenience or loss and without thought of reward, could have evolved. Maybe we worked it out for ourselves rather than having it impressed on us by evolution. If we did, maybe we can work some other things out as well.

Author Lawrence Durrell worked out a variant of Mrs. D's strategy for himself when he concluded from his experience of living in the Greek islands that 'to disarm a Greek, it is only necessary to embrace him'. When he was living in Cyprus just before the outbreak of terrorism that eventually split the country, he had an opportunity to test his conclusion when he was confronted by a belligerent, drunk, knife-wielding neighbour who was muttering imprecations about the presence of the English in his village. Instead of reacting with confrontation, he stepped up and embraced the neighbour, saying, 'Never let it be said that the Greek and the English drew the sword upon each other.' 'No, never,' agreed the surprised neighbour, sheathing his knife and embracing Durrell in return.

Other people do not always respond reciprocally, however. An offer of kindness can be seen as a sign of weakness, as happened to a friend of mine when he let someone stay in his house for a week, only to have them squat in it for the next six months! In a shameful incident from my own days as a student, a fellow student early in the term let me use the water that he was patiently keeping on the boil to heat my own samples. For the rest of the term I continued to use his water, never thinking to take the trouble to boil my own. In the next term he started to pay me back by using my water and not boiling his own. I suppose I asked for his retaliation.

Mrs. B and Mrs. D Get Together

What strategies can we adopt to avoid cycles of retaliation, and yet not become vulnerable to those who would see our efforts to cooperate as weakness and take advantage of us when we offer cooperation? University of Michigan game theorist Robert Axelrod uncovered a stunningly simple answer in 1980, when he invited professionals in his area to submit programs for a Prisoner's Dilemma computer tournament. Pairs of programs played against each other in a game in which they could choose on each move whether to cooperate or whether to defect on the cooperation, basing their decision on what the other program had done in its previous moves. As with all Prisoner's Dilemma situations (whether artificially constructed or arising in real life), the highest rewards went to those who defected when the other had offered cooperation. Mutual cooperation resulted in a somewhat lower reward, but mutual defection was lower still, while the offer of cooperation when the other side defected produced no reward at all (this is known to game theorists as the 'sucker's pay off').

The eight game theorists who submitted programs in response to Axelrod's invitation came up with some ingenious strategies, but when all the programs had eventually played against each other, the winner turned out to be the program that used the simplest strategy of all. It was submitted by the late professor Anatol Rapoport from the University of Toronto, and all that it involved was offering cooperation on the first move, and thereafter echoing whatever the opponent did. It started, in other words, with the gentle Mrs. D strategy but was prepared to follow up with the retributive strategy of Mrs. B if Mrs. D's cooperative approach didn't work.

Axelrod couldn't quite believe that such a simple strategy could be so effective, so he organised a larger tournament, which attracted sixty-two entrants from six countries. Despite the huge variety of strategies on offer, some of them based on the ways in which we handle conflict and cooperation in the real world, the winner was again Rapoport's entry, which he had christened TIT FOR TAT.* Axelrod thought that it might serve as good basic advice to national leaders in their interactions with the leaders of other countries. 'Don't be envious,' he paraphrased. 'Don't be the first to defect, reciprocate both cooperation and defection, and don't be too clever.'

TIT FOR TAT had an enormous impact among social scientists when Axelrod published a popular account of his discoveries in *The Evolution of Cooperation*, because it seemed to offer a neat and simple answer to the problem of cooperation. I thought that I would try it out for myself to see how it worked in an everyday situation, and found my opportunity when a local bookshop had a half-price sale. There were books piled everywhere, with people picking them up, glancing at the titles, and either holding on to them or discarding them. I initiated a cycle of cooperation with the man next to me by showing him the ones that I intended to discard before putting them to one side. Some of these he took for himself, and soon he started to show me the ones that *he* had picked up. In this way we man-

---

*'Tit for Tat' will continue to refer to specific strategies, and 'TIT FOR TAT' will refer to computer programs that embody such strategies.

aged to scan a greater range of books and work our way down the pile quickly. At one stage he stopped showing me what he had picked up, and I immediately reacted by not showing him what I had picked up. He quickly cottoned on that I was reciprocating his defection, and started to cooperate again.

In this case the Tit for Tat strategy seemed to work quite well to initiate and maintain cooperation, but its main value has been to help us think about the problem of cooperation from a new perspective. It has especially been taken up by evolutionary biologists, who have always been puzzled as to how cooperation could have evolved in nature in the face of 'survival of the fittest'. They have discovered a partial answer in Tit for Tat, which does not have to mean retaliation and counter-retaliation, with the strongest eventually coming out on top. It can also mean 'you scratch my back and I'll scratch yours', with evolution favouring those who are most adept at promoting and maintaining cooperation. The ability to cooperate with other members of a group, it seems, has often been a key to survival. In the case of the human race, anthropologists now believe that it has been a major factor, with small cooperative social groups better able to adapt and survive than isolated individuals or groups rent by social schism.

Why Be Nice?

Successful cooperative social groups need their members to be altruistic and cooperative, sacrificing individual advantage for the sake of the group. But why should humans (or other animals) make such a sacrifice? Why should we resist the temptation to cheat for individual advantage, which lies at the heart of the Prisoner's Dilemma? Biologists have discovered one answer

in 'kin selection', in which the evolutionary advantage of cooperating with closely related individuals lies in the preservation
and passing on of one's genetic inheritance. The fierceness with
which a mother tiger will protect her cubs is certainly paralleled
by the fierceness with which many of us will defend our children, but protection of our genetic inheritance does not explain
all of the aspects of human cooperation and social behaviour by
far. Our tendency to play fair in the Ultimatum Game, for example, has nothing to do with whether we are genetically related to
the other players. It might, however, have a lot to do with the
fact that as a species we have somehow become imbued with a
sense of fairness and an ability to empathise with the problems
of others.

It would be nice to think that we could use fairness and empathy to help overcome the various social dilemmas that we encounter. One way to do this is to use Mrs. D's 'do as you would
be done by' strategy, in the hope that others with a similar sense
of fairness and empathy will do the same. We often use Mrs. D's
strategy in families, in relationships, and in the office. The game
theorist's explanation of our cooperative behaviour in these situations is that they involve repeated interactions and that we risk
reprisals if we do not behave with fairness and empathy. One
key to social stability is to behave in a similar way with people
whom we are likely never to meet again. But do we? And why
should we? A number of important social experiments have
been carried out to find answers to these questions.

One of the most interesting experiments was carried out on
students at Princeton Theological Seminary, who were sent by
their teachers to another building to give a talk on the subject
of the Good Samaritan. They did not know that they were the
subjects of an experiment on the parable itself. Their route took

them past an actor who was slumped in a doorway, coughing and obviously in distress. The aim of the experiment was to find out whether thinking about the parable would encourage the students to implement it. The answer? It didn't! The major factor was how much of a rush the students were in. If they were not in a rush, two-thirds of them stopped to help. If they were in a big rush, only 10 per cent helped the 'victim'. The rest did not stop, and some even stepped over him in their rush. Of the forty students on whom the experiment was tried, sixteen offered to help, but twenty-four did not, which the organisers of the experiment took to indicate that private motives can often outweigh public compassion, and that thinking about the subject of compassion makes no difference to our chances of performing the act.

I inadvertently tried a similar experiment for myself when a wheel broke on a heavy suitcase I was carrying on a journey that encompassed several countries. Some people looked away as I struggled; others asked if I needed help. I didn't actually need the help, but I became very interested in how many people were willing to make the offer, so I started to make my struggles more obvious, both in airports and when I was walking along the street. By my count, an average of ten fit and healthy-looking men (I focused on males in this experiment) walked past before one asked if I needed help, with the average being remarkably constant across countries that included Australia, India, England, China, and America.

The fact that even a small proportion of people were willing to behave so altruistically, though, raises the question of just why they do it. One answer is they have inherited a tendency to be altruistic. Another answer is that they have been trained to be altruistic from an early age, to the extent that they would now feel uncomfortable if they did not go out of their way to help others. 'Give me the child until he is seven,' goes the Jesuit saying, 'and I will show you the man.' The essential truth of this saying is undoubted, not just in the matter of religious upbringing (I still carry prejudices from my Methodist upbringing that I have never fully succeeded in shaking off), but in our early cultural upbringing in general. The extent to which our early lives are reflected in our futures is dramatically demonstrated in the British TV series *Seven Up!* (released in the United States as *Age 7 in America*), which follows the lives of a group of children from different social backgrounds, interviewing them every seven years (up to age forty-nine at the moment in the present British series), and finding that their basic paths through life were largely laid down in their early childhood.

I know that I still follow many of the social rules that I was taught in childhood. Mrs. D is still in there somewhere, pushing the buttons to direct operations. There is another drive, though, that keeps us on the straight and narrow for most of the time. It is the existence of social norms. But where do such norms come from, and how does society go about enforcing them? We don't know much about where they come from, but all of the evidence suggests that the enforcer is the retributive Mrs. B.

Social norms are important guidelines for cooperation. They are, in the words of economists Ernst Fehr and Urs Fischbacher, 'standards of behaviour that are based on widely shared beliefs [about] how individual group members ought to behave in a given situation'. But what makes us stick to these standards is another question entirely. The bulk of the evidence suggests that our primary motivation is the fear of sanctions by other members of the group.

Those sanctions can range from disapproval to social exclusion and worse. The most extreme form of social exclusion (short of actual killing) is *ostracism* – a term that originated in ancient Athens, where potential tyrants or people who were thought to be a threat to the state were simply told to go away for ten years. These days the term can cover everything from the little girl who tells her playmates, 'I'm not talking to you', to workers who have been ostracised by their fellow workers for strike-breaking, to people who are suffering from AIDS in Thailand whose lives have been saved by cheaply available antiretroviral drugs but who have nevertheless been ostracised by their families and so are forced to seek refuge in Buddhist temples.

In these cases, the people excluded were known personally to all members of the group from which they were excluded, but this does not always have to be the case. All that is necessary is that the perpetrator or perpetrators be identified by other members of the group. Striking waiters in a New York picket line, for example, carried concealed cameras to photograph union members who had forsaken the strike and threatened to post the photographs at union headquarters so that all members would know the identities of the strike-breakers. In another case, the Chinese American community was alerted to be on the look-out for a man of Chinese origin who had abandoned his three-year-old daughter at an Australian railway station and fled to the United States. Widespread community disapproval meant that the man was quickly identified from his photograph and captured when he attempted to merge anonymously with the Chinese community in Atlanta, Georgia. Mrs. B would surely have approved of the citizens' action in removing his trousers and using them to tie his ankles together until the police arrived.

This incident was an extreme case of *third-party punishment* – that is, punishment inflicted by people who were in no way involved with the original misdemeanour but simply disapproved of it in principle. This sort of punishment is one of the main reinforcers of social norms; it is a way of expressing our disapproval not just on behalf of ourselves but also on behalf of society as a whole. When we turn and glare at someone who is talking at a concert, we do it not just on our own behalf but on behalf of everyone in the audience. Its apotheosis for me occurred in Switzerland, where I saw a tourist drop a sweet wrapper in the street, only to have a local resident pick it up, run after her, and *hand it back*, pointing to a rubbish bin as she did so. The fact that the tourist's face went bright pink was sufficient indication of the efficacy of the action.

Third-party punishment invokes our personal psychology to maintain social norms throughout a community by using our outrage (or at least irritation) that someone has deviated from the norm as a driving force. Laboratory experiments have shown that we are willing to inflict such punishments even at some cost to ourselves. A passive observer watching two other people in a Prisoner's Dilemma-type experiment, for example, will sacrifice real money just for the pleasure of punishing one of the participants who has defected from cooperating with the other. If both players in the constructed game cheat on the cooperation, however, the observer is much less likely to punish either of them. According to psychologists Jeffrey Stevens and Marc Hauser, the authors of the study in which this behaviour was observed, this demonstrated that defection is considered much less of a norm violation if the defection is mutual, while unilateral defection is considered to merit substantial punishment.

Our real-life behaviour reflects that of the participants in the constructed situation. The evidence is that many of our social norms are maintained by *conditional cooperation* – in other words, we will cooperate to maintain the norm (via third-party punishment, for example) so long as most others do, but if too many people break the norm, we feel perfectly free to transgress as well, without much fear of punishment, third-party or otherwise. 'Everyone else is doing it; why shouldn't we?' we say as we dump our rubbish by the side of the road or fudge the returns on our income tax. The eventual result is the collapse of the social norm.

'The social norm of conditional cooperation,' say Stevens and Hauser, 'provides a proximate mechanism behind the famous tit-for-tat strategy.' This is because it involves *indirect reciprocity*, by which Mrs. B's retributional strategy can extend through a community because any individual may perform the retribution for transgression of a social norm on behalf of the community, even though they themselves were not directly affected.

This indirect way of establishing and maintaining social norms is unique to humans, because it requires a combination of psychological ingredients that only we possess. Stevens and Hauser have especially identified numerical quantification (of reward and punishment), time estimation (so that the threat of punishment is not discounted at too high a rate in time), delayed gratification, detection and punishment of cheaters, analysis and recall of reputation, and inhibitory control.

Quite a list! The analysis, transmission, and recall of reputation is especially important. When I go to a new restaurant, for example, the quality of the food and service affects my opinion, but I am unlikely to give the staff direct feedback (although there have been exceptions). I am more likely to pass on my opinion to my friends, and if they go to the restaurant, they will have indirectly reciprocated the good food and service that I received.

I am in effect using Rapoport's Tit for Tat strategy, albeit by paying the restaurant back indirectly rather than directly. This sort of indirect effect can permit cooperation to spread through a community by disseminating the reputations of enough people as cooperators. The problem with using Tit for Tat to do this, however, is that just one defection can produce an endless cycle of tit-for-tat defections, which smacks more of the eternal retribution in Dante's Hell than life in a fair and reasonable world. (The Hell is made even worse in that the endless defection can come from a mistake as well as from a deliberate action.)

New Strategies for Ongoing Cooperation

Can we improve on Tit for Tat as a strategy to maintain and promote cooperation? As it turns out, we can. One new approach was discovered by Martin Nowak and Karl Sigmund when they showed that a strategy of Win-Stay, Lose-Shift does even better than TIT FOR TAT in Axelrod's computer game and that it is closer to the way we often behave in real life. TIT FOR TAT is soulless and unforgiving, as befits its success in a virtual world. Nowak and Sigmund's program, which they call PAVLOV (after the famous Russian scientist who studied conditioned reflexes in animals), uses the Win-Stay, Lose-Shift strategy to model the human attributes of forgiveness and hope. The program continues to cooperate so long as the other program does but will also (unlike TIT FOR TAT) offer cooperation if both programs have lost out through mutual defection in their last encounter, in the hope that the other program might be designed to resume cooperation if cooperation is offered. Technically, say Nowak and Sigmund, it 'embodies an almost reflex-like response to the pay off: it repeats its former move if . . . rewarded, but switches behaviour if . . . punished'.

This was just what my fellow buyer at the book sale did when I defected in response to his defection from our cooperation. By reacting with fresh cooperation, he was unconsciously using the PAVLOV strategy. The originators of the strategy explained its success in the following way:

> The conspicuous success of the tit-for-tat (TFT) strategy . . . relies in part on the clinical neatness of a deterministic cyber-world. In natural populations, errors [and random perturbations] occur . . . [and] occasional mistakes between two TFT players cause long runs of mutual backbiting. (Such mistakes abound in real life; even humans are apt to vent their frustrations upon innocent bystanders.). . .
>
> *Pavlov* has two important advantages over TFT: (1) an inadvertent mistake between two Pavlovians . . . causes one round of mutual defection followed by a return to joint cooperation [and] (2) . . . Pavlov has no qualms about exploiting a sucker. . .
>
> We observe *Pavlov*-type behaviour daily ourselves. Usually, a domestic misunderstanding causes a quarrel, after which cooperation is resumed; and the advice "never give a sucker an even break" is frequently adopted among members of our species.

PAVLOV is just one of many variants of the Tit for Tat strategy that are now being studied extensively. The original Tit for Tat is now classified as a *trigger strategy*, by analogy with gunfights in the Wild West (at least as depicted by Hollywood), in which a shot on one side can provoke the other to respond by pulling the trigger one or more times. Game theorists now recognise a whole range of trigger strategies, all of which follow Mrs. B's rule that non-cooperation will be punished by one or more rounds of non-cooperation in return.

The strongest of these is the Grim Trigger, which threatens, 'If you fail to cooperate even once, I will never, ever cooperate with you again.' The threat of one partner leaving an unhappy marriage after the next fight, never to return, is a Grim Trigger strategy. So, unfortunately, is the threat of nuclear retaliation that still hangs over the world.

A less irrevocable trigger strategy is Generous Tit for Tat, which will respond to cooperation with cooperation but will also sometimes (not always) respond to defection with a further offer of cooperation. A marriage partner might decide to come back after a while, for example, and give their partner a second chance. (If they will come back only when their partner shows definite evidence of change, they are using ordinary Tit for Tat.)

Any of these strategies might succeed. Any of them might fail. Generous Tit for Tat is less punishing than harshly retributive Mrs. B, because it occasionally introduces the forgiving strategy of Mrs. D to break cycles of retribution and counter-retribution. It looks like the best practical approach to many of life's problems. According to relationship psychologists with whom I have discussed the matter, it is the one that is most closely aligned to the psychologically based strategy 'be firm, but be prepared to forgive'. Computer simulations have shown, however, that it is outperformed by PAVLOV, which continues to cooperate so long as the other party does but which will also automatically offer cooperation if both parties have lost out through mutual defection in their last encounter.

I had the chance to test PAVLOV at a cocktail party where a friend and I had both agreed to support each other in going on the wagon in anticipation of driving home. He soon succumbed to the temptation to have a drink, and I thought, 'If he's going to, then I will.' As soon as we saw each other cheat on the co-operation by taking that first drink, we each played PAVLOV by offering not to drink if the other one didn't, and the situation was saved.

Win-Stay, Lose-Shift, by offering cooperation when both parties have lost out through cheating on a previous encounter, seems to be the most effective of all the trigger strategies that have so far been investigated. All of them rely on the power of repeated interactions to induce and maintain cooperation. There is another factor, though, that the memory of previous encounters doesn't even enter into.

The Proximity Factor

The evolution of cooperation isn't just about strategies. When people become neighbours, their geographical proximity must surely count for something in the evolution of their coopera-tion. As it turns out, it counts for a lot. Geographical proximity can produce clusters of cooperators (that is, individuals who use cooperation as their primary strategy), ready to maintain their cooperation in the face of invading defectors. Closing ranks and using cooperation to protect a small group from attack and in-vasion by outsiders happens in villages and small towns. It hap-pens in professional organisations, such as those of doctors and lawyers. It happens in institutions, including some academic institutions where I have worked. And now it is happening in the bowels of computers, where the details of the process are gradually being unravelled by game theorists.

One of their major discoveries has been that cooperation can be maintained by geographical proximity of cooperators even without the memory of how others have behaved in previous encounters, which is a prerequisite for using Tit for Tat strategies. All it needs is two populations, one of cooperators and the other of defectors, that stick with their strategies through thick and thin, just as my mother and nanna did with their respective strategies.

In the first round of one pioneering computer simulation, members of the two populations were arranged randomly on a lattice that looked like a giant chessboard. The individual players were placed on the squares and allowed to interact only with their eight nearest neighbours. The score for each player was the sum of the pay offs from encounters with these neighbours, those pay offs being arranged in the usual Prisoner's Dilemma order. For the next round, the original occupant of each square was retained if he had happened to have the highest score, but was otherwise replaced by the highest scorer among the eight neighbours. It was dead simple and highly revealing. It also made a great video.

The video showed fluctuating patterns emerging as the battle for dominance between cooperators and defectors raged, with clusters of cooperators and clusters of defectors. To the great surprise of the experimenters, neither group wiped out the other. When the dust settled, it turned out that around one-third of the final population consisted of cooperators, while two-thirds consisted of defectors. A proportion of the cheats had prospered, but the cooperators had survived, partly by clustering together in close proximity and partly because the defector's cheating strategy was a losing one when too many cheats got together and mutually sabotaged each other.

Bringing the Threads Together: Repeated Interactions, Proximity, and the Evolution of Cooperation in the Real World

What does all of this mean for the evolution of cooperation in the real world? The studies that I have highlighted provide several strong pointers:

- The effect of geographical proximity means that small communities that rely on mutual cooperation have a much greater chance of maintaining that cooperation than do larger, more dispersed groups, although the computer studies suggest that cheats can survive and prosper to an alarming extent even within such small groups.

- It is much easier to generate cooperation when there is the possibility of one or more repeated encounters between individuals. Burglars, for example, are cheats in the above sense (since they place their individual wants or needs above those of the community), but a number of studies have shown that they are significantly less likely to reoffend if a part of their sentence involves meeting up with and being confronted by their victims.

- In the wider context, the threat of retaliation and retribution can be enough to deter antisocial behaviour and convince people to stick to social norms, especially when that retribution can come not only from the affected parties but also from any member of the social group.

- Reputation is an important incentive, and even embarrassment from being caught breaking a social norm can sometimes be enough. One example that has been studied concerns the incidence of hand-washing by men in public lavatories: men are much more likely to wash their hands if someone else is present, rather than endure the disapproving glance that follows if they do not.

- The most effective strategies for establishing and maintaining cooperation combine elements of Mrs. B and Mrs. D by offering cooperation even when another party has not cooperated but always retaining the option to stop cooperating if the other party does not cooperate. Teddy Roosevelt's 'speak softly and carry a big stick' was one strategy along these lines, but the computer simulations of game theory suggest that it can be more effective to lay greater emphasis on speaking softly, and less on the threat of the stick, by offering cooperation immediately after both sides have defected from the cooperation.

Martin Nowak has recently brought all of these elements together in a wonderful synthesis, 'Five Rules for the Evolution of Cooperation', based on the notion that a cooperator is someone who pays a cost (c) for *another* individual to receive a benefit (b). The individual cooperator loses out, but we know that a *population* of cooperators has a higher average evolutionary fitness (that is, its chance of surviving and reproducing) than does a population of defectors. So what should be the cost-benefit relationship for cooperating if cooperation is to survive and flourish?

Nowak identifies five different mechanisms for the evolution of cooperation, each of which has a different cost-benefit relationship:

1. **Kin Selection:** The *coefficient of relatedness* (which is higher the more closely related the two individuals are) must be greater than the cost-to-benefit ratio (*c:b*).

2. **Repeated Interactions (Direct Reciprocity):** The chance of a future encounter between the same two individuals must be greater than the cost-benefit ratio (*c:b*) of the altruistic act.

3. **Indirect Reciprocity:** This is where our actions are influenced by the effect that they will have on our reputation in the wider community if word gets around. Nowak concludes that indirect reciprocity can only promote cooperation if the probability of knowing someone's reputation is greater than *c:b*.

4. **Network Reciprocity:** This covers the effect of having cooperators or defectors as neighbours, and all that is needed for cooperation is to have a greater number of neighbours than *c:b*.

5. **Group Selection:** A group of cooperators might be more successful than a group of defectors, as happens in the Stag Hunt social dilemma. This case is slightly more complex than the others, because groups will increase in size with time as more offspring are added to the group and may split to form smaller groups. In the mathematically convenient

limit in which selection for cooperation over defection is weak, and groups split only rarely, there is still a surprisingly simple result: cooperation will evolve if the *benefit-cost* ratio (*b:c*) is greater than [1 + ([maximum group size]/[number of groups])].

Nowak's remarkable synthesis shows that we can resolve social dilemmas – and make it possible for cooperation to evolve – if we can get one of the five mechanisms going and if we can find some way to push the benefit-cost ratio in the practical situation above a critical value. It brings many of the strategies I have investigated in this book into a single unifying framework. There is one other strategy for resolving social dilemmas, though, and that is to change the game itself so that the temptation to cheat, which lies at the heart of all social dilemmas, is reduced or eliminated entirely. In the next chapter I investigate some ways in which this might be done, including a remarkable application of the science of quantum mechanics, which can be used in a quite unexpected way to cut the Gordian knot that ties us up in so many social dilemmas.

8

Changing the Game

HOW CAN WE CHANGE THE GAME to improve our chances of cooperation? One way is to introduce new players, which can have quite extraordinary and counter-intuitive consequences. Another way, which will become possible in the near future, is to use quantum computers to set up negotiations that will enable us to read each other's intentions before deciding on our own actions, so that the cheating that lies at the heart of social dilemmas will no longer pay. Here I examine both of these approaches and discover just how they can lead to long-term cooperation.

Introducing New Players

One surprising way to produce harmony and cooperation from conflict, disagreement, and discord is to introduce an even more discordant person into the situation. P. G. Wodehouse's character, the scheming butler Jeeves, describes how this strategy can work in *Right Ho, Jeeves*. 'There is nothing that so satisfactorily unites individuals who have been so unfortunate as to quarrel amongst themselves,' he explains to his long-suffering employer,

Bertie Wooster, 'as a strong mutual dislike for some definite person. In my own family, if I may give a homely illustration, it was a generally accepted axiom that in times of domestic disagreement it was necessary only to invite my Aunt Annie for a visit to heal all breaches.'

When I read this story as a child I was very impressed by Jeeves's advice, and I decided to try it on my parents when they were squabbling over a game of Monopoly – a game that always stirred them to strong passions. I invited the rather grubby boy next door to come in and play, knowing how much my parents disliked his unwholesome presence in their clean and tidy house. Suddenly they were all sweetness and light, and suggested that they should stop their game of Monopoly and take me to the zoo. The boy next door went back home, I got to go to the zoo, and (best of all from my point of view) my parents stopped arguing.

The most discordant people of all are those who enter into competition or conflict with you, but even these can help to stimulate cooperation. Game theorists Peter Fader and John Hauser cite the example of the U.S. microelectronics industries, which were being 'adversely affected by the growing influence and economic power of foreign competition'. Their response was to increase cooperation on basic and applied research, even though the individual firms risked losing their competitive research advantage relative to other U.S. firms.

How does a non-cooperator stimulate cooperation? Fader and Hauser looked for answers in a series of groundbreaking experiments in which they set up a computer tournament similar to that previously organised by Robert Axelrod, except that three computer programs at a time were pitted against each other, rather than just two. The rewards for cooperating or defecting were calculated from a formula that took account of the strategies of all three players and that produced a hierarchy of rewards similar to that for the normal two-person Prisoner's Dilemma (cooperating pays, and solo defecting pays more, but mutual defecting pays less than cooperation). Participants from university groups and major corporations around the world took up the challenge of designing programs whose rules for defecting or cooperating in response to other strategies would maximise their reward from this formula. The game was cast as a marketing game, with price as the sole variable. It was made slightly more complex, but also more realistic, by having a continuum of possible prices that the players could select in order to earn different profits, with collusion and undercutting as the two main strategies. Different degrees of collusion and undercutting were possible depending on the price that each participant selected.

The tournament was run in two rounds, with forty-four entrants in the final mix. The winning program (designed by Australian Bob Marks) was designed to maintain total cooperation if both others cooperated, to defect if both others defected, but to align itself with the program that came closest to a cooperative strategy in other cases by using that same strategy. It was, in other words, designed to look for the best possibilities for cooperation and to exploit them.

Fader and Hauser concluded from the results of the competition that it often pays to be more cooperative in multiperson situations than a simple Tit for Tat strategy would suggest, and that magnanimity and forgiveness are key factors in promoting cooperation in the presence of a non-cooperator. They called the winning strategy *implicit cooperation*. I decided to find out whether people use this sort of strategy in real life by conducting an experiment at a dinner party. When the food was served at the table in large dishes, I ostentatiously helped myself to more than my fair share as the dishes were passed around. A few responded to my strategy by cheating as well, though not to the dramatic extent that I had. When the time came for second helpings, my fellow guests implicitly cooperated by passing the dishes to each other but taking care to bypass me. When the dishes finally came around to me, there was very little food left. My non-cooperative strategy had certainly helped to promote cooperation among the other guests!

My fellow guests did not discuss their strategy; they just followed it, with the assumption that all would do likewise. When we *can* discuss our strategies, cheating by some individuals can also promote cooperation between others, as happened when a spate of thefts occurred in the English village where I live. The presence of burglars, with their defecting strategy, drove us to form a neighbourhood watch scheme, in which we kept more of an eye on each other's properties than we had previously. Our community was more cohesive, and people became more cooperative, after we had banded together to protect ourselves from the burglars.

Cheating is not the only way in which an extra player can help to produce cooperation. Another way that cooperation problems between two parties can be resolved when they don't trust each other is for them to trust a third party. A policeman friend provided me with an unusual example when he told me that he finds it much easier to get miscreants to cooperate and come quietly if he is accompanied by a fierce police dog. The dog is the trusted third party, since both he and the offender can trust it to attack if the offender puts up any resistance!

The more usual approach for ensuring cooperation is for people to post a bond with the third party – something of value that will later be returned so long as both people maintain the cooperation. When I was a student renting a flat it was common practice for both landlords and tenants to post bonds with an independent tribunal. If the landlord didn't maintain the flat, the tenant could complain and ask for repairs to be paid out of the landlord's bond. If the tenant defaulted on the rent, the landlord could claim it from the tenant's bond. In both cases, the tribunal could be trusted to act as an independent and trustworthy third party.

Game theorists have shown that posting a bond can resolve many apparently intractable dilemmas that involve sequential actions. A favourite example is the strangely named Centipede Game, in which a pot of money is passed backwards and forwards between two players a fixed number of times, and they can divide it in half at the end. Each time they pass it, the size of the pot increases. At any stage, however, the player holding the pot can take a substantial proportion, say 60 per cent, of the total for themselves, leaving the other with a lesser amount. It is best for all parties if they keep passing the pot. A logical dilemma analogous to the Prisoner's Dilemma dictates, however, that the person who initially holds the pot should take their 60 per cent and run.

The logic of thinking forward and reasoning backwards reveals the problem. It is the logic that we use in real life when we look forward to predict the consequences of different actions and then reason backwards from the consequences to work out the best action to take. In the case of the Centipede Game, thinking forward shows us that the player who holds the pot last is going to grab their high proportion, rather than divide the final pot 50:50. Reasoning backwards from this point, the person who holds it second-last should grab their high proportion at this stage, rather than passing the pot. But this logic also applies to the person who holds it third-last, and so on all the way back to the beginning, leading us to the conclusion that the person who first holds the pot should take their money and run.

I wondered whether we learn to use this sort of logic as children, and decided to try it out by introducing the Centipede Game at a children's party, using gummy bears rather than money, and adding more gummy bears as the pot was passed from child to child. The wise little eight- to ten-year-olds quickly worked out that they would do better by taking a majority of the pot at the first opportunity, and that was the end of the game.

We can beat the logic that leads us to take a profit at the first opportunity in the Centipede Game by posting a bond. This changes the reward structure so that it is worth continuing the game in order to get your bond back to add to the profits. In the basic game between two players, only *one* player needs to post a high enough bond. That player would then lose out by not passing the pot at any stage, while the other player knows that the one who posted the bond will lose if they defect and so can feel confident about passing the pot on.

I tried the idea of posting a bond with the children at the party by getting each of them to hand over a small present that they had received, with the promise to give it back if the Centipede Game went right through to the end. I was amazed at how quickly they caught on to the idea, and the game did indeed go through to the end.

Some people have argued that the Centipede Game does not reflect real-life scenarios. Others have argued that it reflects the profit-taking strategies of asset-stripping and pork-barrel politics rather well. At the very least, it shows that posting a bond with a trusted third party can make cooperation happen purely on the basis of self-interest.

There is also another way to achieve the same objective without the need for a trusted third party. That is to set up a situation in which each party can see in advance whether the other party intends to cheat or cooperate, and then modify their strategy accordingly. It sounds like an impossible dream, but the surprising and unlikely-sounding combination of game theory and quantum mechanics has made it a realistic possibility and opened up a whole slew of new opportunities for resolving social dilemmas.

Using Quantum Mechanics to Read Each Other's Minds

Quantum game theory takes us into a futuristic world in which the most pressing problems of cooperation simply disappear, or at least become manageable. In this world most of the seven deadly dilemmas are miraculously solved. Cheats no longer prosper, and cooperators win – so long as they negotiate with the aid of a quantum computer.

Quantum computers are the computers of the future. They are still at the experimental stage, but when they become a practical reality (probably in the next decade or so) they will be so fast they will make today's computers look like mechanical adding machines. They will also allow a totally new form of negotiation:

- Participants in a decision-making process in which they can cooperate, defect, or use a mixed strategy enter their decisions by using their consoles to manipulate the state of a quantum object called a *qubit* to represent those decisions. (You don't need to know what qubits actually are, only that they can be used to represent any mixture of strategies. For more detail, see Box 8.1.)

- As soon as one person has registered a decision, everyone else's qubit is affected because of a phenomenon called *entanglement* that is unique to the quantum world (described in Box 8.1). The participants can indirectly detect these changes and respond appropriately by manipulating the states of their own qubits without ever knowing precisely what the others are doing (indeed, there is no communication or information transfer between the parties in the normal sense). Physicist Giles Brassard has called this process 'pseudo-telepathy'. The crucial difference from normal negotiation procedures is that entanglement allows strategies to be coordinated without direct communication.

- The process continues with everyone manipulating their qubits until a joint set of strategies is reached.

- Since cheating in social dilemmas only gives the cheat an advantage if others do not cheat as well, and since the participants can read each other's intended strategies, the incentive to cheat is reduced or eliminated.

- Quantum strategies have been shown to improve our chances of cooperation in all of the main social dilemmas except for Stag Hunt. They can also be used to produce optimal outcomes in a new form of auction called a *quantum auction*.

➤ BOX 8.1

HOW QUANTUM GAME THEORY WORKS

Ordinary computers send and process information in bits that can be in one of two possible states, just as a switch can be in one of two possible states: on or off. These states usually correspond to the numbers 1 or 0, respectively, when it comes to calculations, but they can equally be made to mean *cooperate* or *defect* when it comes to game theory.

Quantum computers use a different sort of bit, called a *qubit* (short for *quantum bit*). Qubits are still at the experimental stage, but they are known to follow the rules of quantum mechanics, in which their state can be set not only to 0 or 1 but also to any mixture of the two (this is called *superposition*). As soon as someone tries to measure their state, however, they mysteriously flip to either a 0 or a 1. The equivalent in game theory is that they can be set not only to *cooperate* or *defect* but also to a mixture of simultaneous cooperation and defection.

If you can't visualise what this means, don't worry. Einstein had problems with it as well. In fact, he thought it was nonsense and tried to disprove it by pointing to some of the ridiculous conclusions it would lead to. One of these (known as the 'Einstein-Podolsky-Rosen paradox') concerns what happens when two spinning electrons are separated. It is central to the way that quantum game theory works.

Electrons (the carriers of electric current) are often used to realise the concept of the qubit in practice. They have a property called *spin* that turns them into little magnets that can be oriented either 'up', 'down', or (in the mysterious world of quantum mechanics) a mixture of the two. Only when someone tries to measure the spin does this mixture collapse to give either up or down.

The fun starts when two electrons are very close together, in which case quantum mechanics says that their spins must be opposite. An electron only acquires a definite value for its spin,

➤

however, when someone tries to measure it. As soon as you measure one, the other immediately flips to the opposite. If the one that you are measuring comes out as up, the other automatically becomes down.

Einstein thought that this was crazy and asked what would happen if the two electrons were separated and taken to opposite sides of the galaxy, both with their spin still in the indeterminate state that exists before someone tries to measure it. In his seminal paper with Podolsky and Rosen, he argued in effect that if someone on one side of the galaxy tried to measure the spin of one electron, it is ridiculous to think that this would trigger the spin of its distant partner immediately to assume the opposite value.

Amazingly, Einstein and his co-authors were wrong. Experiments have shown that when two such electrons are separated, measuring the spin of one *does* set the spin of its distant partner to the opposite state. This phenomenon (now known as *entanglement*) is what makes quantum game theory (and quantum computers) possible. Game theorists have proved that it could help us to beat social dilemmas and reach truly cooperative decisions and strategies. One of the reasons for this, according to pioneering quantum game theorist Jens Eisert, is that entanglement means that there can be *no* Nash equilibria in pure strategies. In other words, the principal temptation to defect for individual gain (that is, the presence of a Nash equilibrium) is simply not there in these cases. More generally, entanglement permits people to coordinate their strategies without ever directly knowing what the strategies of the other players are.

The way entanglement is used is to entangle two or more qubits (one for each participant) and then to separate them and allow each participant to manipulate the state of his or her own qubit to reflect their decision on whether to cooperate, defect, or use a mixed strategy in a given situation. As soon as one qubit's state has been manipulated, the states of the other entangled qubits are automatically affected. It's as though each participant has a card with *cooperate* written on one side and *cheat* written

➤

on the other. Both cards are initially set on edge, but as soon as one participant tips his or her card over to announce a definite strategy instead of being indeterminate, the other participants' cards automatically tip to the opposite strategy, providing a tip-off about the first person's intentions. The others can then flip their cards in response, which of course provides further tip-offs about different participants' willingness to cooperate.

The net effect is that cheating doesn't pay in many social dilemmas, because cheating only pays if others use a cooperative strategy, and this is not the best strategy if they know for sure that the other will cheat. Manipulation of the qubits is thus likely to lead to a more cooperative outcome because it amounts to pseudo-telepathy, which physicist Tad Hogg says 'allows individuals to pre-commit to agreements' – thus overcoming a major barrier to the resolution of social dilemmas.

Could quantum game theory work in practice? A group of scientists at the Hewlett-Packard laboratories decided to find out by studying its application to the Free Rider problem. A free rider is one who, recognising that he or she cannot be excluded from consuming a good, has no incentive to offer to purchase it. If no one paid up, however, the resource would not exist. Participants in the experiment (students from Stanford University) were placed in the Free Rider dilemma when each was given a certain amount of virtual money and asked to choose how much to invest in a public fund. The total investment was then multiplied by a factor that represented the return on the investment, and the total benefit was distributed equally among the participants, who had been told that this would happen and who had been asked to choose a strategy that would yield them the maximum return. Most of them succumbed to the temptation to cheat, which game theorists have shown to be the dominant strategy and the public fund rapidly dwindled to almost nothing.

The experiment was then repeated with entanglement, which allowed the participants to automatically receive tip-offs about the intentions of the other players and to adjust their own strategies to suit. Entanglement was mimicked by means of a computer program. Each participant was given a 'particle' that could be set to one of two states – invest or don't invest. The trick was that the particles were entangled, so as soon as someone had made a choice, it would affect the states of the other particles, permitting them to adjust their strategies in turn to achieve the best outcome for themselves. In effect, entanglement allowed them to at least partially coordinate their strategies. The overall result was that the participants cooperated roughly 50 per cent of the time, as opposed to 33 per cent, which was all they could achieve without the aid of quantum strategies. Popular-science writer Mark Buchanan describes what happens in this quantum scenario: 'It becomes likely that cheating by one person will be met by cheating by others. [Since all participants know this in advance, they will know that] cheating doesn't pay, and [so] quantum theory deters freeloading and promotes a better outcome.'

The Hewlett-Packard scientists found that the more participants there were, the more they tended to cooperate rather than cheat. 'If confirmed with larger groups,' said science reporter Navroz Patel, 'this effect would be highly desirable in the context of Internet piracy, where the number of players, that is downloaders, can run into tens of millions.'

There are practical barriers, of course, to the implementation of quantum game theory. One is that the development of quantum computers is still in its infancy. Scientists now know how to manufacture and manipulate a few qubits at a time, but creating a working computer that will need thousands or even millions of qubits is not yet a practical possibility. When such computers do become available, the participating parties would still have to recognise their possibilities for making negotiated cooperative agreements easier to attain. One likely place for this to happen initially is in the commercial arena, where businesses of all kinds face the Free Rider dilemma. For example, many small shareholders might benefit from a change in company management or policy, but if a small group of them put the work in to effect the change, all the others will benefit without having done any of the work. The net result is often that no one is willing to do the work, and inefficiencies continue. Another example would be a company that wants to pay lawyers to argue a case for tax breaks on its product deciding not to because then all firms manufacturing similar products benefit from the tax breaks without contributing to the effort.

In these and many other cases, negotiation to share costs with others would ultimately benefit all – if most could be persuaded to cooperate. The results of the Hewlett-Packard experiment suggest that the new negotiation strategies made possible by the use of quantum computers can provide a significantly higher chance of achieving effective cooperation in such situations, as well as in areas such as negotiation for wages and relations between employers and employees. If these possibilities come to fruition, they will be a considerable advance (and also a proving ground) in the search for more effective negotiating strategies for efficient cooperation. As author Adrian Cho has pointed out, the use of entanglement might even drive traders to cooperate, and thus help to create a more crash-resistant stock market.

Quantum game theory is for the future. Classical game theory is for the present, and it provides a rich range of strategies to help us overcome social dilemmas. In the final chapter I briefly review these strategies and come up with a top ten list of tips for strategies that we can use to promote cooperation in our own lives and in the wider world.

Conclusion

Individuals *Can* Make a Difference: The Top Ten Tips

I BEGAN MY INVESTIGATION of game theory with the conviction that we need new strategies to address the social dilemmas that we face every day, both in our personal lives and in the global arena. I ended it by concluding that game theory *can* add to our chances of resolving such problems in two main ways:

1. by helping us to view them from a new perspective that exposes their true underlying causes, and

2. by providing new strategies to help us resolve them.

This is not to say that game theory provides complete answers, but it provides strategies that can often help to tip the delicate balance between cooperation and conflict. Everyone deserves to know about these strategies and how they can use them. Here is my personal selection of the most useful – my top ten tips for cooperation in everyday life:

1. *Stay if you win, shift if you lose.* If your choice between cooperating and using an independent, non-cooperative strategy turns out to be a winner, stick with it. If it doesn't (often because the other person has defected from cooperation at the same time that you have), switch to your other choice of strategy right away.

2. *Bring an extra player in.* If you are involved in a two-way game, turn it into a three-way game. It works for the balance of nature; it can work for the balance of cooperation. It can also pay to bring a player in who you *know* will be a non-cooperator. Finally, the extra player may be able to act as a trusted third party to hold a bond or enforce a contract.

3. *Set up some form of reciprocity.* One of the most important incentives for cooperation is knowing that you will have to interact with the other party again in the future. Try to set up such situations directly, indirectly, or by the creation of social networks.

4. *Restrict your own future options so that you will lose out if you defect on cooperation.* This is one of the most powerful ways of showing another person or group that your commitment to cooperation with them is credible. Examples include putting yourself (or others) in a position in which your reputation or theirs will suffer if you or they do not deliver, and burning your bridges so that you cannot renege on cooperation once it has been agreed upon.

5. ***Offer trust.*** This is another way of offering credible commitment. If you genuinely offer trust, trust will often be returned, making cooperation that much easier.

6. ***Create a situation that neither party can independently escape from without loss.*** This is, of course, a Nash equilibrium. If the cooperative solution to a dilemma is also a Nash equilibrium, your problems are solved.

7. ***Use side payments to create and maintain cooperative coalitions.*** The side payment can be money, social or emotional rewards, or even outright bribery. All that matters is to ensure that people will lose out if they leave your coalition to join or form another one.

8. ***Be aware of the seven deadly dilemmas, and try to reorganise the benefits and costs to different players so that the dilemma disappears.*** This is, of course, not as easy as it sounds, or the world would be a happier place. It is a step in the right direction, though, and always worth a try.

9. ***Divide goods, responsibilities, jobs, and penalties so that the result is envy-free.*** Our sense of fairness is a strong motivator; use it by setting up situations in which the process is agreed upon and transparent, and the outcome is obviously fair.

10. ***Divide large groups into smaller ones.*** I have deliberately left this very important strategy until last. All of the evidence points to the fact that cooperation is much easier to

engender in small groups, but the downside is that cooperation *between* such groups becomes more difficult. This dichotomy lies at the heart of many of the serious problems that I outlined at the start of this book. The tips above could help if group leaders or representatives used them to promote cooperative coalitions of such groups. It happens when families and small social groups get together to form larger communities. Maybe it could be made to happen on a wider scale. One would certainly hope so.

Some of these strategies might seem like no more than common sense, but game theory adds extra dimensions by showing just why and how they work in different circumstances. Some can seem quite counter-intuitive, and these have emerged directly from game theory itself. It should also be recognised that they are only a start. Game theory is a young science; it is advancing rapidly now, and we will continue to see advances in the years to come. One direction in which those advances are already occurring is through the use of *complexity theory*, which deals with complex systems, such as society as a whole, rather than breaking them down into smaller units (such as two-person interactions) that are easier to think about and analyse. Complexity theory is now beginning to address some of the social dilemmas identified by game theory. Another future approach will be to use the uncertainties inherent in quantum theory to produce more certainty in our attempts to cooperate. This is what happens when some molecules spontaneously get together to form cooperative units in living cells. By understanding how this sort of spontaneous cooperation happens, we may be able to get a better handle on how to make it happen in our own society. I began this book because I was worried about the problems that society now faces and wanted to understand what game theory had to offer in terms of strategies for cooperation. Here I have shared my journey of discovery, and the prospects that the strategies of game theory offer. I hope that the next time you look at a newspaper or watch a television programme you'll scream 'game theory!' (either out loud or at least to yourself), and that you will enjoy seeing and using game theory in action in your own life. Thank you for sharing the journey.

Notes

All URLs are current as of March 2010.

Introduction

2   ***there is another side to game theory – a side that concerns cooperation*** Regrettably, it is a side that often goes unacknowl-
edged. When Professor Jeffrey Sachs, director of the Earth
Institute at Columbia University and a former special advisor
to the United Nations, delivered the 2007 BBC Reith Lectures
on the idea that the Earth is 'Bursting at the Seams' (www.bbc
.co.uk/radio4/reith2007/), he spoke a great deal about the need
for cooperation but never once mentioned the Tragedy of the
Commons, game theory, or any of the concrete strategies that
have emerged from game theory.

2   ***hidden barrier to cooperation*** The barrier arises from an underlying paradox in logic, but this is not to say that emotional
issues are unimportant for the problems of cooperation – quite
the reverse. Psychologist Daniel Goleman makes the point that
'feelings are typically indispensable for rational decisions; [their
function is to] point us in the proper direction, where dry logic
can then be of best use' (*Emotional Intelligence* [London: Blooms-
bury Publishing, 1996], 28).

   Sigmund Freud made a related point when he said that 'society has had to enforce from without rules meant to subdue tides
of emotional excess that surge too freely within' (paraphrase of
Sigmund Freud, Civilization and its Discontents, in Goleman,
Emotional Intelligence, 5).

2   ***catch-22 logical trap*** Joseph Heller's *Catch-22* was first published in 1961 by Simon & Schuster. The now-famous eponymous term refers to any situation where circular logic catches the victim in an inescapable double bind.

3   ***planet entirely populated by spoon life-forms*** This idea possibly derived from Douglas Adams's *Hitchhiker's Guide to the Galaxy*, where there was 'a planet entirely given over to biro life forms. And it was to this planet that unattended biros would make their way, slipping away quietly through wormholes in space to a world where they knew they could enjoy a uniquely biroid lifestyle . . . and generally leading the biro equivalent of the good life' (Douglas Adams, The Hitchhiker's Guide to the Galaxy [London: Pan Books, 1979], 113).

3   ***example of the Tragedy of the Commons*** Garrett Hardin, 'The Tragedy of the Commons,' *Science* 162 (1968): 1243–48. The full essay is available at dieoff.org/page95.htm. The parable of a group of herders grazing their animals on common land was originally introduced by William Forster Lloyd in his book *Two Lectures on the Checks to Population* (Oxford: Oxford University Press, 1833).

3   ***scientists applied the same argument to teaspoons*** Megan S. C. Lim, Margaret E. Hellard, and Campbell H. Aitken, "The Case of the Disappearing Teaspoons: Longitudinal Cohort Study of the Displacement of Teaspoons in an Australian Research Institute,' *British Medical Journal* 331 (2005): 1498–1500.

5   ***'Everybody's crying peace on earth, Just as soon as we win this war'*** Mose Allison, 'Everybody's Cryin Mercy' (1968). First released on album *I've Been Doin' Some Thinkin'* (Atlantic Records, 1968: Cat. No. SD1511).

5   ***we are not all Mother Teresas*** My wife was very struck by a radio interview with Mother Teresa in which she said that she did not act out of altruism but to satisfy an inner, personal, and ultimately selfish need (driven, she believed, by God).

5   ***'Stern Review on the Economics of Climate Change'*** www.hm-treasury.gov.uk/sternreview_index.htm.

5   *Game theory . . . accepts the fact that self-interest is one of our primary motivations* The comic strip Doonesbury puts a nicely cynical spin on it in a wonderful exchange between Zonker Harris and Kirby, a new boy at Walden University who is seeking Zonker's advice on how to understand the seventies:

ZONKER: They say you can always tell a culture by its literature! Well, we've got just about every movie novelization and self-help manual published in the last ten years! . . . Personally, I favor the output of the new school of amorality. Looking out for you-know-who just seems so sensible these days!

KIRBY: Gee, I dunno, Zonk. I'm not sure that's me . . .

ZONKER: You? Who cares about you?

KIRBY: Oh, wow . . . you really sound in control of your life!

Garry Trudeau, *The People's Doonesbury: Notes from Underfoot, 1978–1980* (New York: Holt, Rinehart, and Winston, 1981).

6   *the film* A Beautiful Mind was based on the book of the same name by Sylvia Nasar (New York: Simon & Schuster, 1998). The book gives a good basic account of Nash's contribution to game theory; the film gives little, and what little it gives it unfortunately mangles.

6   **mamihlapinatapai** Michelle McCarthy and Mark Young, eds. *Guinness Book of Records* (New York: Facts on File, 1992).

7   **Robert Axelrod's book, The Evolution of Cooperation** New York: Basic Books, 1984.

7   *a later foreword by the biologist Richard Dawkins* London: Penguin, 1990.

Chapter 1

13   *Puccini's opera* Tosca The person who first spotted the appropriateness of their predicament for game theory was Anotol Rapoport in 'The Use and Misuse of Game Theory', *Scientific American* 207 (1962): 108–118.

Anecdotes abound about how real-life performances of *Tosca* have also been plagued by social dilemmas. According to one story, the lead soprano had been treating the stagehands in a very high-handed way during rehearsals instead of cooperating with them to produce the best overall performance, and the stage-hands took their revenge by removing some of the mattresses on which she was to land behind the stage after she plummeted from the castle. She is supposed to have broken an ankle, and they lost their jobs.

Over-cooperation also has its perils. Author Gerald Durrell recounts the story of a performance on the Greek island of Corfu where the overenthusiastic stagehands provided rather too many mattresses behind the stage for the plummeting soprano to land on, with the result that her upper parts reappeared several times to the view of the mystified audience.

14 ***Princeton University mathematician Albert Tucker*** The full history of this now-famous story is given in Poundstone, *Prisoner's Dilemma*. Albert Tucker was John Nash's Ph.D. supervisor.

15 ***price-fixing by supermarkets*** *The Independent*, December 9, 2007.

16 ***the history of the Dead Sea Scrolls*** The financial aspect of the scrolls was not confined to Bedouin shepherds. The June 1, 1954, issue of the *Wall Street Journal* contained an advertisement reading, 'The Four Dead Sea Scrolls: Biblical manuscripts dating back to at least 200 BC are for sale. This would be an ideal gift to an educational institution or religious institution by an individual or group' (see Ayala Sussman and Ruth Peled, 'The Dead Sea Scrolls', *Scrolls from the Dead Sea* [Washington, D.C.: Library of Congress, 1993], www.jewishvirtuallibrary.org/jsource/History/deadsea.html).

17 ***the whole field of ethics . . . comes down to . . . historical attempts to get around the . . . Prisoner's dilemmas and other social*** See, for example, J. L. Mackie, *Ethics: Inventing Right and Wrong* (New York: Penguin, 1991). Science-fiction writer Isaac Asimov, inventor of the Three Rules of Robotics, makes the interesting point that

> if you stop to think about it, the three Rules of Robotics are the essential guiding principles of a good many of the world's ethical systems. Of course, every human being is supposed to have the instinct of self-preservation. That's Rule Three to a robot. Also every 'good' human being, with a social conscience and a sense of

responsibility, is supposed to defer to proper authority; to listen to his doctor, his boss, his government, his psychiatrist, his fellow man; to obey laws to follow rules, to conform to custom – even when they interfere with his comfort or his safety. That's Rule Two to a robot. Also, every 'good' human being is supposed to love others as himself, protect his fellow man, risk his life to save another. That's Rule One to a robot.' (Isaac Asimov, *The Complete Robot* [London: HarperCollins, 1982], 530).

17 **'Mathematicians are comparatively sane'** *New Scientist*, December 18, 2004, 46.

18 **'This man is a genius'** Professor R. J. Duffin, Carnegie Institute of Technology. Story related in Harold W. Kuhn, 'The Work of John Nash in Game Theory', Nobel Prize seminar, December 8, 1994, nobelprize.org/nobel_prizes/economics/laureates/1994/nash-lecture.pdf.

22 **one of the shortest scientific papers ever to win . . . a Nobel Prize** 'Equilibrium Points in N-person Games', *Proceedings of the National Academy of Sciences* 36 (1950): 48–49. These are probably the most important two pages of socially important mathematics in history, although some writers like to point out that the French economist and mathematician Antoine Augustin Cournot discovered a version of Nash's theory in his famous (to economists) 'duopoly' model of business competition. Cournot did not, however, go on to prove its ubiquitous occurrence in our society.

The prize is actually the Sveriges Riksbank Prize in Economic Sciences in Memory of Alfred Nobel. It was awarded in 1994 to John Harsanyi, John Nash, and Richard Selten for their 'pioneering analysis of equilibria in the theory of non-cooperative games'. Nash has often been quoted as saying that it was for his 'most trivial work'. A video of a 1994 interview with Nash on the effect of the prize on his life can be seen at nobelprize.org/nobel_prizes/economics/laureates/1994/nash-interview.html.

23 **paradoxical circle of logic** This sort of paradox has been known at least since the time of the Cretan philosopher Epimenides, who lived in the sixth century B.C.E. and made the famous assertion, 'All Cretans are liars.' (Think about it; was Epimenides himself a liar according to this statement?) A modern version is

a card that has printed on one side 'The statement on the other side of this card is false', while on the other side is printed 'The statement on the other side of this card is true.'

23 *Too often, parties will agree to a negotiated compromise and then one party will break the agreement when it suits them* This is exactly what Adolf Hitler did when he signed the Munich Agreement with Neville Chamberlain, Benito Mussolini, and douard Daladier in September 1938. The agreement handed de facto control of Czechoslovakia to Germany. (It should not be confused with the abortive England–Germany peace treaty that was later signed by Hitler and Chamberlain alone.) The three non-German signatories attempted to minimise the possibility of war by permitting Germany's annexation of Czechoslovakia, so long as Hitler agreed to go no further. Hitler beat their strategy by agreeing to the deal and then breaking the bargain and invading Poland a year later, when he had had time to build up Germany's military strength.

24 *senior Church of England bishop* Bishop Peter Price, the Bishop of Bath and Wells.

25 *'evolution . . . has solved the problem for . . . ants, bees, and wasps by genetically programming them to cooperate'* The multitudinous authors of the article announcing the sequencing of the honey bee genome say that 'most mysteries of sociality appear to be encoded subtly in [it]' (Honey Bee Genome Sequencing Consortium, 'Insights into Social Insects from the Genome of the Honeybee *Apis mellifera*', *Nature* 443 [2006]: 931–49).

With regard to the human genome, biologist Richard Dawkins says that "each [gene is] selected for its capacity to cooperate with the others that it is likely to meet" (interview on *The Science Show*, ABC [Australia], April 22, 2006, www.abc.net.au/rn/scienceshow/stories/2006/1617982.htm).

Prior to our knowledge of the genome, many parallels were drawn between our own societies and those of the ants. I particularly enjoy two of them, both from Caryl P. Haskins, Of Ants and Men (London: Allen and Unwin, 1945), 69, 99:

> A study of ant societies [is] a criterion of the background of the guiding forces which have molded the complex basis of our own social structure.

and:

> When we compare the motives which bind together the societies of humans and ants, we are forcibly struck by their similarities. Fundamentally, of course, the purpose of social organization is precisely the same in both creatures – to promote individual welfare and security, to permit the individual to live more peaceably in his immediate environment and to reproduce with greater safety, and to obtain that margin of social security which will provide for his needs in time of famine and uncertainty. . . . Individuals of both groups labor under a force which may well be called 'social pressure'.

Perhaps Isaac Asimov should have the last word. In *Robots and Empire*, robots have begun to modify the human race so as to protect it from itself, arguing that 'we must shape a desirable species and then protect it, rather than finding ourselves forced to select among two or more undesirabilities' (London: Grafton Books, 1986, 465).

25   *ridiculous for us to . . . rely on nature* A biologist colleague with a macabre sense of humour has suggested that one end result of such evolution could be the division of the human species into two races, with one keeping the other as a meat animal, as the Morlocks did with the Eloi in H. G. Wells's *The Time Machine*.

  Some evolutionists argue that the human race has stopped evolving, but these arguments seem to be based on political correctness as much as they are on sound science (see, for example, Kate Douglas, 'Are We Still Evolving?' *New Scientist*, March 11, 2006, 30). My personal view is that we are subject to increasing selection pressures and that we are bound to evolve, although hopefully not according to philosopher Dan Dennett's pessimistic vision: 'Perhaps we will so befoul our planet that only an eccentric and hardy remnant of our species – which can survive on earthworms while living in underground burrows, for instance – will remain' (quoted in Douglas, 'Are We Still Evolving?').

26   *Plato's idea* Plato made his suggestion in Book VII of *The Republic*. His selected trainees were supposed to undergo two years of rigorous physical training and then study mathematics for ten years, which would surely have finished off most modern

kings and presidents. Then they had to follow a fifteen-year apprenticeship under the guidance of a philosopher before taking up their kingship, which they had to share around with other philosopher-kings.

Interestingly, Ayatollah Khomeini was so impressed by Plato's notion that he aimed to model his Islamic Republic on Plato's pattern.

26 **King Solomon** certainly knew how to handle a situation. When two women appeared before him, each claiming motherhood of the same baby, he called for a sword so that he could cut the baby in two and give them half each. It was quite a cunning ploy to identify the real mother, because he realized that she would be the more likely to give up her claim rather than have her baby killed (1 Kings 3:16–18).

26 **His yearly take of gold alone** 1 Kings 10:14.

26 **£39 billion that was left to him to build his famous temple** 1 Chronicles 22:14.

28 **the law . . . can even be an ass** This comes from Charles Dickens, *Oliver Twist*, chapter 51: "'If the law supposes that,' said Mr. Bumble . . . "the law is a ass – a idiot. If that's the eye of the law, the law is a bachelor; and the worst I wish the law is that his eye may be opened by experience – by experience.'"

28 **United Nations Core International Human Rights Treaties** Office of the United Nations High Commission for Human Rights, 'The New Core International Human Rights Treaties' (New York and Geneva: United Nations, 2007), www.ohchr.org/Documents/Publications/newCoreTreatiesen.pdf.

Chapter 2

33 **a sense of justice** Philosophers from ancient Greek times to the present day have been preoccupied with what it means to have a just society, with the implicit (and reasonable) assumption that this is what we should all desire. One of the most interesting contemporary analyses is by the Harvard philosopher John Rawls, who defined it as the sort of society that we would choose to be born into if we were forced to choose from behind a 'veil of igno-

rance', not knowing what race or gender we would have, what sort of parents we would have, or even whether we would be born with more or less intelligence, drive, or other personal characteristics (*A Theory of Justice* [Cambridge, Mass.: Belknap Press, 1971]).

33   B*rown capuchin monkeys get frustrated and angry* Sarah F. Brosnan, 'Nonhuman Species' Reactions to Inequity and Their Implications for Fairness', *Journal of Social Justice* 19 (2006): 153–85.

36   *division of property in divorce cases* See, for example, Will Hively, 'Dividing the Spoils', *Discover Magazine*, March 1995 (see www.colorado.edu/education/DMP/dividing_spoils.html).

36   *1994 United Nations Convention on the Law of the Sea* The convention applies only to the deep ocean bed, since a UN treaty states that countries have exclusive mining rights to their own continental shelves. This presents an interesting problem in the Arctic Ocean, where the continental shelves of Russia, Canada, Denmark, Norway, and the United States are all linked via a 'phallic piece of submarine geography' that runs right under the Arctic ice cap, called the Lomonosov Ridge. (The graphic description of its shape first appeared in 'Editorial: Save the Arctic Ocean for Wildlife and Science', *New Scientist*, September 1, 2007, 5.) The first four countries are presently appealing to the United Nations for the right to drill for oil in this area, claiming it to be an extension of their own continental shelves. At the time of this writing, the United States cannot join in the appeal because it has never ratified the treaty.

   For an interesting discussion of the general principle in international law, see Abbas Raza, 'Cake Theory and Sri Lanka's President', *3 Quarks Daily*, April 11, 2005, 3quarksdaily.blogs.com/3quarksdaily/2005/04/3qd_monday_musi.html.

36   *he got Minimax down to a T* Simon had never actually heard of game theory; he arrived at his description by intuition. *Why You Lose at Bridge* (New York: Simon & Schuster, 1946), 3.

37   **Theory of Games and Economic Behavior** John von Neumann and Oskar Morgenstern, 3rd ed. (Princeton: Princeton University Press, 1957).

38   *a fellow guest, who promptly took the* smaller *of the two pieces*
It wouldn't have happened in the North of England – not in the
house of the cartoon character Andy Capp, anyway. The rationale given by this unrepentantly self-centred male chauvinist after he has taken the larger piece is marvellous:

FLO (Andy's long-suffering wife): If that 'ad been me I'd 'ave
'ad the politeness to take the smaller piece.

ANDY: Well, you've got it then, 'aven't you.

There could, of course, be other reasons for taking the
smaller piece. The person concerned could be on a diet. In another country, it could be polite to take the larger piece in order
to show appreciation. None of these invalidates my main point,
which is the difficulty of assessing the overall benefits of an action to a person.

40   *cash value* These have even been attached to body parts. Following the devastating 2004 tsunami in the Indian Ocean, for example, women from the poor village of Eranavoor in southern India
assigned a cash value to their kidneys by selling them for around
£650 each to make ends meet after the family's fishing livelihoods
were destroyed (Randeep Ramesh, 'Indian Tsunami Victims Sold
Their Kidneys to Survive', *The Guardian*, January 18, 2007, www.
guardian.co.uk/world/2007/jan/18/india.tsunami2004).

Interestingly, the American renal transplant surgeon Dr.
Arthur Matas, a former president of the American Society of
Transplant Surgeons, has proposed that the law be changed in
the United States to make the sale of kidneys legal because he
is worried about the shortage of donor kidneys. It has been es-
timated that a healthy kidney would fetch $60,000–$70,000,
which, together with the cost of a transplant operation, would
still be less than the cost of long-term dialysis ('Organ Transplant Expert Answers Our Viewers' Questions About Kidney Sales', *ABC News,* November 22, 2007, abcnews.go.com/
WN/story?id=3902508&page=1).

Even a person's cultural heritage can have a cash value. When
I was visiting Laos a few years ago, I fell into conversation with a
local guide and made the point that Laotians seemed to be hell-
bent on pursuing a better material way of life and losing their
traditional way of life in the process. He laughed in my face. 'You

should try living it,' he said, 'and then see whether you come up with the same argument.'

He was quite right. I was valuing his cultural heritage from my perspective and not from his. But in both cases there was a cash value that could be attached. How much would I, as a representative Western tourist, be willing to pay him to maintain his heritage so that I could enjoy it from the outside? How much would he have to be paid to maintain that heritage and not try to modernise it (with the complication that he might well accept the money and then modernise it anyway)?

The Center for International Forestry Research is now trying out a new method when it comes to resources in Borneo, where they are finding the sort of decision-guiding information that simply doesn't emerge from classical biodiversity surveys (Charlie Pye-Smith, 'Biodiversity: A New Perspective', *New Scientist*, December 10, 2005, 50–53). Researchers ask the indigenous people which resources are most important to them and respect their views when it comes to 'cake cutting' and different uses.

40   **What really worked was a bribe** Bribery has received uniformly bad press in the West, but other cultures can have quite different attitudes toward it. When Westerners try to do business with those cultures, problems arise, which is why the Organisation for Economic Co-operation and Development has recently prepared a policy for 'preventing bribery of public officials' in 'selected Middle East and North African countries' (OECD, 'Business Ethics and Anti-Bribery Policies in Selected Middle East and North African Countries 2006', www.oecd.org/dataoecd/56/63/36086689.pdf).

My personal opinion is that it might be more effective, and more respectful of cultural differences, for countries in the West to prepare a policy on 'accepting bribery in selected Middle East and North African countries'.

40   **Economist Ignacio Palacio-Huerta watched over a thousand penalty kicks** 'Professionals Play Minimax', *Review of Economic Studies* 70 (2003): 395–415.

41   **the farmers would have to be paid to stop digging up their hedges** The answer is around £1 per yard, or up to £9 per yard to

plant a new hedge (e.g., www.durhamlandscape.info/landscape/ usp.nsf/pws/Durham+Landscape+-+Modern+Landscape+- +Field+Boundaries+-+Background+Paper). Until a loophole in the legislation was stopped, this differential tempted many farmers to grub up ancient hedgerows and then replace them with new ones!

43   *utils* This unit of measure lets game theorists construct prefer-
      ence scales. There is a large literature on such scales (e.g., www.
      changingminds.org/explanations/preferences/preferences.htm)
      with the emphasis on comparing preferences for one particular
      thing between different individuals. This is difficult enough, but
      game theorists face an even more difficult task in trying to com-
      pare and add up the preferences of one individual for two or more
      quite different things, which is why they invented the util.

45   *the* **cake-cutting problem** This is discussed in user-friendly
      detail in Jack Robertson and William Webb, *Cake-Cutting Algo-
      rithms: Be Fair If You Can* (Natick, Mass.: A. K. Peters, 1998).

45   *a man who had three wives* This problem is discussed in tech-
      nical detail by Robert J. Aumann and Michael Maschler, 'Game
      Theoretic Analysis of a Bankruptcy Problem from the Talmud',
      Journal of Economic Theory 36 (1985): 195–213. Professor
      Aumann has also produced an excellent nontechnical (and non-
      mathematical) summary of the argument in 'Game Theory in the
      Talmud', Research Bulletin Series on Jewish Law and Economics
      (Toronto: York University, n.d.), dept.econ.yorku.ca/~jros/docs/
      AumannGame.pdf.

      Dollar values of ancient Jewish currencies are discussed in
      Micael Broyde and Jonathon Reiss, 'The Ketubah in America: Its
      Value in Dollars, Its Significance in *Halacha* and Its Enforceabil-
      ity in Secular Law', www.jlaw.com/Articles/KETUBAH.pdf.

47   ***Equal Division of the Contested Sum . . . settling territorial dis-
      putes*** See, for example, Herschel I. Grossman, 'Fifty-four Forty or
      Fight', Brown University Economics Working Paper No. 03-10,
      April 2003, papers.ssrn.com/s013/papers.cfm?abstract_id=399
      781 (abstract).

48   ***Steven Brams*** is a former president of the Peace Science Society
      and has continued to be prolific in producing ideas about power

and cooperation in numerous areas, from influence and power in terrorist networks to a new way of electing committees (see politics.as.nyu.edu/object/stevenbrams.html).

48 *equally applicable to birthday parties and legal parties* Paraphrase of a headline in the *New York Times* (Sarah Boxer, 'For Birthday Parties or Legal Parties; Dividing Things Fairly Is Not Always a Piece of Cake', August 7, 1999).

48 **The Win-Win Solution: Guaranteeing Fair Shares for Everyone** Steven J. Brams and Alan D. Taylor (New York: Norton, 1999).

49 *considerable progress . . . in working out more fair and equitable approaches* See, for example, S. Mansoob Murshad, 'Indivisibility, Fairness, Farsightedness and Their Implications for Security', United Nations University Research Paper 2006/28, March 2006.

51 **Delphi technique** Developed by the RAND Corporation in the 1960s (where game theory was also being developed and exploited), the Delphi technique has been used by many large organisations – not only businesses and government but also organisations like the National Cancer Institute. It is named after the Greek oracle at Delphi. Her name was Pythia, and she was a priestess of the god Apollo. She is supposed to have delivered prophecies inspired by Apollo, but, more likely, they were actually inspired by ethylene gas leaking into the cave where she lived.

For a good description of the modern Delphi technique, see Allan Cline, 'Prioritization Process Using Delphi Technique', white paper, Carolla Development, 2000, www.carolla.com/wp-delph.htm.

51 *averaged opinion of a mass of equally expert or equally ignorant observers* Eric S. Raymond, *The Cathedral and the Bazaar: Musings on Linux and Open Source by an Accidental Revolutionary*, rev. ed. (Cambridge, Mass.: O'Reilly, 2001).

52 *'every week, group intelligence won'* James Surowiecki, *The Wisdom of Crowds: Why the Many Are Smarter Than the Few and How Collective Wisdom Shapes Business, Economies, Societies, and Nations* (New York: Doubleday, 2004).

Chapter 3

56 **simultaneous games . . . *as matrices*** This way of representation was originated by John von Neumann. It is called the 'normal form', as distinct from the branching tree of strategies and outcomes that is the only realistic alternative. Game theorists use both, depending on which is more convenient for the purpose. Matrices are generally used if the strategic decisions (moves) are simultaneous, while the branching decision tree is used if the decisions are sequential and taken alternately by each participant. Von Neumann proved mathematically that the two methods of representation are equivalent.

Even for the simple case where there are just two players, each with a choice between two strategies, the number of possible permutations for the rewards means that there are seventy-eight possible matrices (Melvin J. Guyer and Anatol Rapoport, 'A Taxonomy of 2 × 2 Games', *General Systems* 11 [1966]: 203–14). Most of these are either trivial or benign; only a few trap us in social dilemmas rather like the way the human race was trapped in a virtual reality world in *The Matrix*. In that case, sentient machines were to blame; in the real world, we have only ourselves to blame.

59 ***The Tragedy of the Commons is essentially a multiperson Prisoner's Dilemma*** Game theorists have proved mathematically that the Tragedy of the Commons is equivalent to a series of Prisoner's Dilemmas between individual pairs of people.

61 ***even disarmament*** This argument has been forcefully presented by Jeffrey Rogers Hummel, 'National Goods Versus Public Goods: Defense, Disarmament, and Free Riders', *Review of Austrian Economics* 4 (1990): 88–122, www.mises.org/journals/rae/pdf/rae4_1_4.pdf.

62 ***care and use of communal resources*** The Greek philosopher Aristotle was one of the first to point out the existence of the problem when he observed, 'That which is common to the greatest number has the least care bestowed upon it' (*Politics*, Book II, chapter 3, 1261b, translated by Benjamin Jowett, *The Politics of Aristotle: Translated into English with Introduction, Marginal Analysis, Essays, Notes and Indices* [Oxford: Clarendon Press,

1885]). For a modern version, see www.gutenberg.org/etext/
6762.

64 *free steam heating* I believe that, in some cases at least, the ra-
diators could not be turned off, so the citizens *had* to open the
doors and windows to keep their house temperatures down!

64 *thin-walled apartments* According to Professor Robert E. Marks
of the Australian Graduate School of Management, the communist regime in Hungary followed Karl Marx's belief that accommodation was a means to an end (that of getting workers
into the factories) and did not invest more than was necessary in
accommodation, because this would take resources away from
the factories ('Rising Legal Costs', in *Justice in the Twenty-First
Century*, ed. Russell Fox [London: Cavendish Publishing, 1999],
227–35).

65 *costs to the rest of us would soon mount up* The philosopher
Bertrand Russell was well aware of this when he wrote his
tongue-in-cheek essay 'In Praise of Idleness', but he made a valuable point when he said that 'the idea that the poor should have
leisure has always been shocking to the rich' (*In Praise of Idleness
and Other Essays* [New York: Routledge, 2004], www.zpub.com/
notes/idle.html).

65 *Edward Gibbon* The author of *The History of the Decline and Fall
of the Roman Empire*, Gibbon spent fourteen months as a student
at Oxford and later said that they were 'the most idle and un-
profitable' of his life.

65 *'The working class can kiss me arse'* From Dorothy Hewett *This
Old Man Comes Rolling Home* (play). Sydney: Currency Press,
1976.

67 *the individual official who takes a bribe* According to a report
in the German newspaper *Der Tagesspiegel* (December 17, 1996),
'The Thai Deputy Minister of the Interior, Mr. Pairoj Lohsoon-
thorn, has publicly called on officials to accept bribes. He had
ordered staff of the land sales department of his ministry to accept any money offered to them, he told *Matichon* newspaper.
However, civil servants were not allowed to ask for bribes or to
circulate price lists. "This is part of traditional Thai culture," Mr.

Pairoj said. The acceptance of bribes was justified by the low level of pay in the civil service.'

67   **Vermilion Gate** New York: Abacus, 2002.

68   *Naval commander Gaurav Aggarwal gives a wonderful example* 'The Naval Salute', *Quarterdeck* 18 (2005): www.bharat-rakshak.com/NAVY/Articles/Article07.html.

69   *Cuban Missile Crisis* The widespread interpretation of this event as a game of Chicken is reinforced by Secretary of State Dean Rusk's often-quoted remark, 'We're eyeball to eyeball, and I think the other fellow just blinked.' (See Steven J. Brams, 'Game Theory and the Cuban Missile Crisis', *Plus* [January 2001]: plus.maths.org/issue13/features/brams/index.html for an interesting alternative interpretation of the crisis.)

71   **Common Sense and Nuclear Warfare** London: Allen and Unwin, 1959, 30.

72   *Hawk-Dove* These strategies were first described this way in 'The Logic of Animal Conflict' (J. Maynard Smith and G. R. Price, *Nature* 246 [1973]: 15–18). It seems at first glance that Hawks should always come out on top, so animals using the Dove strategy should be quickly eliminated from the gene pool. It depends on how often individuals using different strategies come across each other. When two Hawks come across each other, they are likely to fight and both get injured; but when two Doves come across each other, the one that runs away more slowly will end up with the spoils, and the other one will at least get away uninjured. The net result is that most animal populations have a mixture of individuals using the different strategies.

72   **katydids** Also known as long-horned grasshoppers, or bush crickets.

74   *overweight people were photographed wearing tiny bathing suits* 'Lose the Weight, or Wear the Bikini on TV', ABC News, March 15, 2006, abcnews.go.com/Primetime/story?id=1725982. Negotiated threats can be difficult to enforce if the threatened party is in a position to renegotiate to stop the threat from being carried out. This particular threat gained added credibility be-

cause it had been advertised, making it difficult for the individuals to renegotiate if they failed to lose weight.

74 *Hern·n CortÈs . . . destroyed the ships* Legend has it that he burned them, but in fact he simply scuttled them after removing the artillery (Winston A. Reynolds, 'The Burning Ships of Hern·n Cortés', *Hispania* 42 [1959]: 317–24).

75 *communication is the key to negotiating coordinated strategies* Many animals can communicate through gesture and display, and this can be seen as a form of negotiation, albeit without the flexibility provided by language.

75 **The Volunteer's Dilemma** This dilemma was accurately identified and described by Aristotle when he said, 'everybody is more inclined to neglect the duty which he expects another to fulfill' ('Politics', Book II, chapter 3, 1261b, translated by Benjamin Jowett, *The Politics of Aristotle: Translated into English with Introduction, Marginal Analysis, Essays, Notes and Indices* [Oxford: Clarendon Press, 1885]). For a modern version, see www.gutenberg.org/etext/6762.

76 **'What if everybody felt that way?'** Joseph Heller, *Catch-22* (New York: Simon & Schuster, 1961).

76 *Yag·n Indians of Tierra del Fuego* The last of the pure Yag·n men, Felipe, died of old age in May 1977.

78 *the answer lies in a heavy hint* According to C. P. Snow 'you couldn't read the diaries of the Scott expedition without realizing that it had been hinted, more than once, to Captain Oates that he ought to go' (*Last Things* [London: Penguin Books, 1972], 310).

78 **Staff Sergeant Laszlo Rabel** was awarded the U.S. Medal of Honor for this act of 'conspicuous gallantry', which took place at 1000 hours on November 13, 1968, in Binh Dinh Province, Republic of Vietnam (www.homeofheroes.com/moh/citations _1960_vn/rabel_laszlo.html).

79 *invitation . . . to send a card asking for either $20 or $100* This can also be seen as a multiperson Prisoner's Dilemma. The

results are analysed more fully in William Poundstone, *Prisoner's Dilemma* (Oxford: Oxford University Press, 1993), 203–4.

79 ***Thomas Schelling*** Schelling's groundbreaking book, *The Strategy of Conflict*, published in 1960 by Harvard University Press, is still a delightful, informative, and thought-provoking read.

80 ***Margaret Thatcher was famous for giving false clues*** Geoffrey W. Beattie, 'Turn-Taking and Interruption During Political Interviews: Margaret Thatcher and Jim Callaghan Compared and Contrasted', *Semiotica* 39 (1982): 93–114.

84 ***Darwin . . . propose marriage to his cousin Emma*** The effort reportedly left him with such a headache, and left Emma so startled by the unexpected proposal, that 'we both looked very dismal' (letter from Emma Wedgwood to Jesse Sisimondi, November 15, 1838, see 'A Wife, That Most Interesting Specimen' [Auckland: Auckland Museum, 2008], http://www.amnh.org/exhibitions/darwin/idea/wife.php). So dismal, in fact, that several aunts who were hanging around hoping to celebrate the engagement went off to bed, assuming that the proposal had somehow misfired.

85 ***Robert Aumann*** According to the Nobel committee publicity release:

> In many real-world situations, cooperation may be easier to sustain in a long-term relationship than in a single encounter. Analyses of short-run games are, thus, often too restrictive. Robert Aumann was the first to conduct a full-fledged formal analysis of so-called infinitely repeated games. His research identified exactly what outcomes can be upheld over time in long-run relations.
>
> The theory of repeated games enhances our understanding of the prerequisites for cooperation: Why it is more difficult when there are many participants, when they interact infrequently, when interaction is likely to be broken off, when the time horizon is short or when others' actions cannot be clearly observed. Insights into these issues help explain economic conflicts such as price wars and trade wars, as well as why some communities are more successful than others in managing common-pool resources. The repeated-

games approach clarifies the raison d'être of many institutions, ranging from merchant guilds and organized crime to wage negotiations and international trade agreements (nobelprize.org/nobel _prizes/economics/laureates/2005/press.html).

86 *Stag Hunt . . . , which game theorist Brian Skyrms believes to be more relevant to the problem of social cooperation than the Prisoner's Dilemma* Brian Skyrms, Presidential Address to the Pacific Division of the American Philosophical Association, March 2001, www.lps.uci.edu/home/fac-staff/faculty/skyrms/ StagHunt.pdf. Professor Skyrms has elaborated his argument, with many interesting examples, in *The Stag Hunt and the Evolution of Social Structure* (Cambridge: Cambridge University Press, 2004).

86 *The name comes from a story told by . . . Jean-Jacques Rousseau* A Discourse on a Subject Proposed by the Academy of Dijon: What Is the Origin of Inequality Among Men, and Is It Authorised by Natural Law? (1754), translated by G. D. H. Cole, and rendered into HTML and text by Jon Roland, www .constitution.org/jjr/ineq.txt.

Jean-Jacques Rousseau produced the statement, 'Man was born free, but he is everywhere in chains.' It appears at the start of his book *The Social Contract* (translated by Maurice Cranston [New York: Penguin, 1968, 1]). It is a frustrating book that offended practically everyone in its day and that is in fact very Machiavellian in its content. He argues, for example, that religion should be the servant of the state and that it should teach patriotic, civic, and martial virtues. He even proposes the death penalty for anyone whose conduct is at variance with that taught by the state religion. So much for individual freedom.

88 *1989 proposed constitutional amendment* William Poundstone, *Prisoner's Dilemma* (Oxford: Oxford University Press, 1993), 220.

Chapter 4

91 *George Washington is reputed to have played it* The origin of this often-quoted story is obscure. Washington and Rochambeau (together with the French Comte deBarras) certainly signed the articles of capitulation for the allies, as did Cornwallis and Thomas Symonds for the British, but there is no mention of exactly *where* they were signed, in Washington's diaries or elsewhere. A transcription of the diaries with detailed notes has been produced by Donald Jackson and Dorothy Twohig, eds., *The Diaries of George Washington. 3: The Papers of George Washington*. (Charlottesville: University Press of Virginia, 1978).

91 *Florida judge ordered two attorneys to play it* Judge Gregory A. Presnell of Orlando ordered this unusual measure after two Tampa attorneys proved unable to agree upon where to hold a deposition, even though both of their offices were just four floors away in the very same building in Tampa ('Order of the Court', *CNNMoney.com/Fortune*, June 7, 2006, money.cnn.com/2006/06/07/magazines/fortune/judgerps_fortune/index.htm.

92 *Most of us know the simple rules* They are not so simple, however, when it comes to the World Championships (yes, they do exist!), which use an elaborate arrangement consisting of prime, approach, and delivery. According to the World Rock Paper Scissors Society's website:

> The Prime is the ritual used to get players in sync with each other so they can deliver their throws simultaneously. It is the action of retracting one's fist from full-arm extension towards the shoulder and then back to full extension. This phase is critically important. If at any time the players are not in sync with their primes, then play must stop and begin again. Having players deliver their throws at the same time is critical to ensuring a fair match.
>
> Priming conventions generally fall into two classes:
>
> 1) European Prime: Three prime shoot. Players pump their arms in unison three times before starting the Approach phase.
>
> 2) North American Prime: Two prime shoot. Players pump their arms in unison twice before starting the Approach phase.

> The Approach is the transition phase between the final prime and the Delivery. As one's arm makes its final descent a player is required to make a decision about the throw they will make. The Approach begins at the shoulder following the final prime and ends when the arm makes a 90-degree angle with the player's body. Players must reveal their chosen throw to their opponent prior to reaching the 90-degree mark. Any throw delivered past this critical point must be considered a Forced Rock (since this is the position the hand would have been in upon crossing the 90-degree mark).

It's all designed for drama rather than fairness. One obvious way to get a fair result would be for the opponents to stand back to back, holding their hands behind them, and for an independent referee to declare a winner or a draw after each had formed their throws. But it wouldn't be so much fun.

93 *Christie's and Sotheby's auction . . . Impressionist paintings* Carol Vogel, 'Rock, Paper, Payoff: Child's Play Wins Auction House an Art Sale', *New York Times*, April 29, 2005, www.nytimes.com/2005/04/29/arts/design/29scis.html.
Auction houses give code names to their sales. Christie's called this one Scissors, for fairly obvious reasons.

93 *episode of* **The Simpsons** Episode 9F16, 'The Front', FOX, 1993, written by Adam I. Lapidus and directed by Richard Moore.

93 *Scissors tends to be played slightly less often in tournaments* The figure 29.6 per cent is given on the World RPS Society website, www.worldrps.com.

94 **intransitive** *nature of Rock, Paper, Scissors* As a refresher, a transitive relationship in logic and mathematics means that if A is greater than B and B is greater than C, then A must be greater than C. This is the normal state of affairs with things like numbers, heights, speeds, and whether one object is behind another, etc. *Intransitivity* is much rarer, and can be a bit of a puzzle until you stop to think about it.

95 *Californian side-blotched lizard* Kelly R. Zamudio and Barry Sinervo, 'Polygyny, Mate-Guarding, and Posthumous Fertilization as Alternative Male Mating Strategies', *Proceedings of the Na-*

tional Academy of Sciences (U.S.), December 2000, www.pnas.org/cgi/content/abstract/011544998v1 (abstract).

96 **biodiversity in bacterial neighbourhoods** Benjamin Kerr, Margaret A. Riley, Marcus W. Feldman, and Brendan J. M. Bohannan, 'Local Dispersal Promotes Biodiversity in a Real-Life Game of Rock-Paper-Scissors', *Nature* 418 (July 11, 2002): 171–74.

96 *self-enforcing balance . . . an important component of biodiversity* Richard A. Lankau and Sharon Y. Strauss, 'Mutual Feedbacks Maintain Both Genetic and Species Diversity in a Plant Community', Science 317 (September 2007): 1561–63). See also Richard A. Lankau, 'Biodiversity: A "Rock-Paper-Scissors" Game Played at Multiple Scales', *Scitizen*, scitizen.com/screens/blogPage/viewBlog/sw_viewBlog.php?idTheme=22&idContribution=1076.

98 *a Loner or Volunteer strategy* C. Hauert, S. De Monte, J. Hofbauer, and K. Sigmund, 'Volunteering as Red Queen Mechanism for Cooperation in Public Goods Games', Science 296 (2002): 1129–32, and 'Replicator Dynamics for Public Goods Games', *Journal of Theoretical Biology* 218 (2002): 187–94.

98 *'volunteering relaxes the social dilemma'. . . Manfred Milinski and his group* Dirk Semmann, Hans-J.rgen Krambeck, and Manfred Milinski, 'Volunteering Leads to Rock-Paper-Scissors Dynamics in a Public Goods Game', *Nature* 425 (2003): 390–93.

101 *truel* The word was coined by Yale University economist Martin Shubik in the 1960s (see Kilgour/Brams paper below).

102 *Marc Kilgour and Steven Brams . . . who first analysed the truel* 'The Truel', *Mathematics Magazine* 70 (1997): 315–16.

103 *both houses of Parliament have had to be dissolved seven times* In 1972, 1976, 1979, 1983, 1994, 1996, and 2008.

104 *Roshambot* Perry Friedman, personal communication, February 20, 2008.

Chapter 5

107 **Herrings . . . communicate by farting** You can listen for your-self (you predator, you) at news.nationalgeographic.com/news/2005/11/1118_051118_herring_video.html ("Video in the News: Do Herrings Fart to Communicate?" *National Geographic*, November 18, 2005).

108 **Clive James . . . interposing a noisy, gaseous commentary** The incident is described in his autobiography, *Unreliable Memoirs* (London: Picador, 1981). This book comes with a health warning: people look at you in a peculiar way when you start laughing uncontrollably while you are reading it in public, as I did when I read that 'two bacon rolls and a custard pie were my undoing'. The book also reflects far too many details of my own Australian childhood that I would rather no one else knew anything about.

108 **Le Pétomane** was the stage name of Joseph Pujol. In addition to farting La Marseillaise, he could play a flute with his backside and blow out a candle from several yards away. The climax of his act was a noisy impression of the 1906 San Francisco earthquake. A musical based on his life, called *The Fartiste*, was awarded Best Musical at the 2006 New York International Fringe Festival, www.broadwayworld.com/viewcolumn.cfm?colid=15679.

108 **'waggle dance'** See http://www.youtube.com/watch?v=4Nteg AOQpSs for a good video and description of the science involved. See also Thomas D. Seeley, *The Wisdom of the Hive: The Social Physiology of Honey Bee Colonies* (Cambridge: Harvard University Press, 1996).

108 **Farting, dancing, and laying down odour trails** The synchronisation of menstrual cycles in all-female communities, such as those in a convent, is driven by unconscious odour cues, called the 'McClintock effect' (see Martha McClintock, 'Menstrual Synchrony and Suppression', *Nature* 229 [1971]: 244–45, and K. Stern and M. K. McClintock, 'Regulation of Ovulation by Human Pheromones', *Nature* 392 [1998]: 177–79). McClintock's group and others have since used 'sweaty T-shirt' experiments, in which women sniffed T-shirts that had been worn by men to show that they use aroma cues to prefer partners whose genetic background is different from their own (see Suma Jacob, Martha

J. McClintock, Bethanne Zelano, and Carole Ober, 'Paternally Inherited HLA Alleles Are Associated with Women's Choice of Male Odor', Nature Genetics 30 [2002]: 174–79).

108 ***humpbacks produce songs that have a hierarchical syntax*** R. Suzuki, J. R. Buck, and P. L. Tyack, 'Information Entropy of Humpback Whale Songs', *Journal of the Acoustical Society of America* 119 (2006): 1849–66. The authors used statistics and the science of information theory to show that there is a complex, non-random pattern in the sounds that the whales produce. To hear those sounds, go to www.newscientist.com/article/dn8886-whale-song-reveals-sophisticated-language-skills.html and click on the indicated link (Roxanne Khamsi, 'Whale Song Reveals Sophisticated Language Skills', NewScientist.com, March 23, 2006).

A good summary of this work was given by journalist David Baron in a talk, 'Information Theory and Whale Song', *The Science Show*, ABC Radio, Australia, June 17, 2000, www.abc.net.au/rn/science/ss/stories/s140922.htm.

109 ***bit is the smallest piece of information that allows a distinction between two possibilities*** A switch, for example, can be off or on, which, in computers, corresponds to a 0 or a 1.

109 ***President Kennedy . . . holds the world record of 327 words per minute*** This is the record for a speech made in public life, as listed in the *Guinness Book of Records*. The speech in question was made in December 1961, at the end of the first year of his presidency.

109 ***5.5 bits per phoneme*** This way of characterising language was pioneered by E. Colin Cherry, Morris Halle, and Roman Jakobson in 'Toward the Logical Description of Languages in Their Phonemic Aspect', *Language* 29 (1953): 34–46. It has its origins in the work of the irascible English philologist Henry Sweet, who developed a shorthand way of writing down words as a set of phonemes and who was the origin of Professor Henry Higgins's character in George Bernard Shaw's play *Pygmalion* (later turned into the musical My Fair Lady).

109 ***4 to 6 phonemes per word*** It depends on the complexity of the words that you use, of course. It would be towards the lower end

for everyday speech, towards the upper end for a scientific conference talk, and probably off the scale for a discussion between experts at a game theory conference. See the review by Noam Chomsky, 'Langage des machines et langage humain', *Language* 34 (1958): 99–105.

109 ***Negotiation*** I am, of course, only scratching the surface and focusing mainly on negotiation between two individuals or groups of individuals. Extra strategies are available when many people are involved. For example, as I show later in the chapter, when one person is negotiating with many, it can actually pay to reduce the value of the assets that are being offered! See Adam M. Brandenburger and Barry J. Nalebuff, *Co-opetition* (London: HarperCollins, 1996).

109 ***'the one who produces the most excrement is usually the eventual winner'*** William R. Hartston and Jill Dawson, *The Ultimate Irrelevant Encyclopaedia* (London: Unwin Paperbacks, 1985), 102. Bill writes the humorous 'Beachcomber' column for the *Daily Express*, and one suspects that his description is slightly slanted in the typical exaggerated 'Beachcomber' style.

110 ***a dark band appears between the closely spaced fingers*** Full details are given in my book *Weighing the Soul: Scientific Discovery from the Brilliant to the Bizarre* (New York: Arcade, 2004).

111 ***George Melly . . . discovered a threat that was really effective*** This incident is described in the first volume of Melly's autobiography, *Owning Up* (Harmondsworth: Penguin, 1970).

113 ***'any nonconstant sum game can, in principle, be converted to a win-win game'*** Roger A. McCain, Game Theory: A Non-technical Introduction to the Analysis of Strategy (Mason, Ohio: Thomson/South-Western, 2004), 183. The shift in mindset is quite a major one. Game theorists distinguish between non-cooperative game theory (in which the participants are not in a position to negotiate effectively) and cooperative game theory (in which negotiating enforceable agreements is possible and inducements to cooperate can be offered [e.g., side payments or threats of retaliation]).

114 ***They were caught in a Prisoner's Dilemma*** The easiest way to see this is to take Frank and Bernard back to their childhoods

and look at the expressions on their faces when they either keep their gifts or give their gifts to the other:

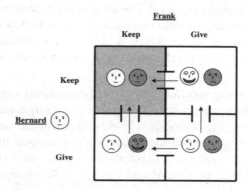

They would both be better off with give-give (bottom right) but they can't get there unless they form a coalition and coordinate their strategies because keep-keep is *dominant* (just follow the arrows) if each acts in his own individual interest independent of the other (for a fuller explanation, see Box 3.1).

115 **back-stabbing, gossiping, and switching of allegiances** A particularly clear example may be found in C. P. Snow's *The Masters* (New York: Charles Scribner's Sons, 1951).

116 **two oarsmen . . . sitting side by side** Hume tells this story in *Treatise on Human Nature* (Book III, part 2, section 3 [New York: Oxford University Press, 2000]).

117 **driven to form a coalition by mutual self-interest** As the Canadian philosopher David Gauthier puts it, 'each man has two possible actions, to row, or not to row. Each prefers the outcome if both row . . . to the outcome if neither rows, or indeed, to any other possible outcome. [They have arrived at] a dominant, stable convention . . . which serves to coordinate [their] actions [so that] their preferences converge on the choice of a mode of behavior and on adherence to the mode chosen' ('David Hume, Contractarian', *Philosophical Review* 88, no. 1 [January 1979]: 3–38).

117 **minimally effective cooperation** In the general, multiparticipant case, game theorists describe what the members of the various

coalitions get after all the side payments have been made as an *allocation*. The allocation is *efficient* if there is no way to rearrange these allocations so that one or more people are better off without making anyone else worse off. The set of coalition structures that corresponds to this condition is called the *core*. Paraphrased from Roger A. McCain, *Game Theory: A Non-technical Introduction to the Analysis of Strategy* (Mason, Ohio: Thomson/South-Western, 2004), 185.

119 **Nash bargaining solution** This is a Pareto optimal solution to the bargaining game described by Nash ('The Bargaining Problem', *Econometrica* 18 [1950]: 155–62). The ramifications of the four conditions listed in the main text are discussed in Shaun Heap and Yanis Varoukis's wonderful little book *Game Theory: A Critical Introduction* (London and New York: Routledge, 1995), 118–28.

120 **Negotiations for the purchase of advertising media time** Leonard Greenhalgh and Scott A. Neslin, 'Nash's Theory of Cooperative Games as a Predictor of the Outcomes of Buyer-Seller Negotiations: An Experiment in Media Purchasing', Journal of *Marketing Research* 20 (1983): 368–79, and Scott A. Neslin and Leonard Greenhalgh, 'The Ability of Nash's Theory of Cooperative Games to Predict the Outcomes of Buyer-Seller Negotiations: A Dyad-Level Test', *Management Science* 32 (1986): 480–98.

121 **'the greatest auction ever'** William Safire, 'The Greatest Auction Ever', *New York Times*, March 16, 1995, A17.

121 **spectrum auctions** The Federal Communications Commission describes them as follows:

> In a simultaneous multiple-round (SMR) auction, all licenses are available for bidding throughout the entire auction, thus the term 'simultaneous'. Unlike most auctions in which bidding is continuous, SMR auctions have discrete, successive rounds, with the length of each round announced in advance by the Commission.
>
> After each round closes, round results are processed and made public. Only then do bidders learn about the bids placed by other bidders. This provides information about the value of the licenses to all bidders and increases the likelihood that the licenses will be assigned to the bidders who value them the most.

The period between auction rounds also allows bidders to take stock of, and perhaps adjust, their bidding strategies. In an SMR auction, there is no preset number of rounds. Bidding continues, round after round, until a round occurs in which all bidder activity ceases. That round becomes the closing round of the auction. (FCC, 'Simultaneous Multiple-round [SMR] Auctions', August 9, 2006, wireless.fcc.gov/auctions/default.htm?job=about_auctions &page=2.)

121 *'the trouble with game theory is that it can explain anything'* Richard Rumelt, discussion comment at the conference 'Fundamental Issues in Strategy: A Research Agenda for the 1990s', (Richard P. Rumelt, Dan E. Schendel, and David J. Teece, eds. [Boston: Harvard Business School Press, 1994]).

121 *Rumelt's Flaming Trousers Conjecture* 'Burning Your Britches Behind You: Can Policy Scholars Bank on Game Theory?' *Strategic Management Journal* 12 (1991): 153–55.

123 **The Front** The film is made more powerful by the fact that its writer and director, along with many of the actors (including Mostel), were themselves blacklisted during that era.

123 *Indonesia, . . . where offers of $30 or less were frequently rejected* Lisa Cameron, 'Raising the Stakes in the Ultimatum Game: Experimental Evidence from Indonesia', *Working Paper #345*, Industrial Relations Section, Princeton University, 1995, quoted in Robert Slonim and Alvin E. Roth, 'Learning in High-Stakes Ultimatum Games: An Experiment in the Slovak Republic', *Econometrica* 66, no. 3 (May 1998): 569–96.

123 *watch . . . the brains of the participants* Alan G. Sanfey, James K. Rilling, Jessica A. Aronson, Leigh E. Nystrom, and Jonathan D. Cohen, 'The Neural Basis of Economic Decision Making in the Ultimatum Game', *Science* 300 (2003): 1755–58.

124 *the Ultimatum Game . . . is 'catching up with the Prisoner's Dilemma'* Martin A. Nowak, Karen Page, and Karl Sigmund, 'Fairness Versus Reason in the Ultimatum Game', Science 289 (2000): 1773–75.

Chapter 6

127 **'If you can't trust dogs and little babies'** Originally published on October 25, 1959. *Don't Give Up, Charlie Brown* (Greenwich, Ct.: Fawcett Publications Inc., 1974).

127 **Sir Walter Raleigh reputedly took off his cloak** This story is actually pure fiction, probably originated by Thomas Fuller in his book *Anglorum Speculum or the Worthies of England* (1684). It was further propagated by the novelist Sir Walter Scott in his 1821 Elizabethan romance *Kenilworth* when the queen says: 'Hark ye, Master Raleigh, see thou fail not to wear thy muddy cloak.'

128 **Lucy van Pelt, Charlie Brown . . . the football** The example is from *Don't Give Up, Charlie Brown* (Greenwich, Ct.: Fawcett Publications Inc., 1974). Cartoonist Charles Schulz wrote in 1976 that 'Lucy has been inviting Charlie Brown to come running up to kick the football and then pulling it away each year for eighteen years. Every time I complete this annual page, I am sure I will never be able to think of another one, but so far I have always managed to come up with a new twist for the finish. . . . I was told by a professional football player that he actually saw it happen in a college game at the University of Minnesota' (*Peanuts Jubilee: My Life and Art with Charlie Brown and Others* [London: Penguin, 1976], 91).

In fact, Schulz managed to keep the theme going for twenty years. Some of Lucy's other comments were:

'I'm a changed person. . . . Isn't this a face you can trust?' (1957)

'You have to learn to be trusting.' (1959)

Charlie Brown pulls back at the last moment, expecting to catch Lucy in the act. 'Don't you trust anyone any more?' He then tries again, with the predictable result. (1961)

'A woman's handshake is not legally binding.' (1963)

In 1976, she even tells him that she is going to pull it away, but he takes no notice. 'Men never really listen to what women are saying, do they?'

129 **trust performs three functions** Barbara Misztal, *Trust in Modern Societies: The Search for the Bases of Social Order* (Cambridge: Polity Press, 1996).

129 **any non-constant sum game . . . can, in principle, be converted to a win-win game** Roger A. McCain, *Game Theory: A Non-Technical Introduction to the Analysis of Strategy* (Mason, Ohio: Thomson/South-Western, 2004), 183.

130 **we face a crisis in the first year of our lives** Erik H. Erikson *Childhood and Society* (New York: Norton, 1963). Erikson adds reliable physical comfort to the reliable emotional comforts listed in the main text.

130 **measure the effect of the change on a person's willingness to trust** Michael Kosfield, Markus Heinrichs, Paul Zak, Urs Fischbacher, and Ernst Fehr, 'Oxytocin Increases Trust in Humans', *Nature* 435 (2005): 673–76. It is still too early to say whether the childhood crisis postulated by Erikson affects our ability to manufacture this hormone.

131 **having their subjects play a trust game** Similar trust games, played in many different laboratories around the world, have now convinced most game theorists that we are often more altruistic, sharing, and caring than pure self-interest would seem to dictate.

132 **You can't get trust in a bottle** You can, however, put it in your pocket. Australian Commonwealth Scientific and Research Organization scientist John Zic and his team have developed a 'trust extension device' that can be plugged into strange computers. Carried on a memory stick or a mobile phone, it makes trust portable, opening the way for secure transactions to be undertaken anywhere, even in an Internet cafÈ. The device creates its own environment on an untrusted computer and, before it runs an application, it establishes trust with the remote enterprise server. Both ends must prove their identities to each other and also prove that the computing environments are as expected. Once the parties prove to each other that they are trustworthy, the device accesses the remote server and the transaction takes place.

132   *our brains, which some scientists argue have evolved in two parallel ways* Nobel laureate Vernon L. Smith, for example, focused his prize acceptance speech on 'the simultaneous existence of two rational orders'. Smith himself was largely responsible for identifying our duality of thinking when it comes to economic matters through his pioneering of 'experimental economics', the results of which have shown that the self-seeking, self-centred *Homo economicus* of game theory and classical economics is largely a myth, and that fairness is a stronger motivation than many would have guessed (see *Papers in Experimental Economics* [Cambridge: Cambridge University Press, 1991], and *Bargaining and Market Behavior: Essays in Experimental Economics* [Cambridge: Cambridge University Press, 2001]).

    Smith suffers from Asperger's Syndrome, which makes it harder for him to read the non-verbal cues that most of us take for granted, and which would seem to be a distinct problem for someone trying to understand non-rational human interactions. He says, however, that it helps. 'I can switch out and go into a concentrated mode and the world is completely shut out,' he said in a recent interview. 'If I'm writing something, nothing else exists.' 'Perhaps even more importantly, I don't have any trouble thinking outside the box. I don't feel any social pressure to do things the way other people are doing them' ('Mild Autism Has "Selective Advantages": Asperger Syndrome Can Improve Concentration', interview by Sue Herera, CNBC, February 25, 2005, www.msnbc.msn.com/id/7030731/).

132   *Machiavellian intelligence* The controversial Machiavellian intelligence hypothesis (also known as the 'social brain hypothesis')

> identifies selective forces resulting from social competitive interactions as the most important factor in the evolution of hominids, who at some point in the past became an ecologically dominant species. These forces selected for more and more effective strategies of achieving social success (including deception, manipulation, alliance formation, exploitation of the expertise of others, etc.) and for ability to learn and use them. The social success translated into reproductive success selecting for larger and more complex brains. Once a tool for inventing, learning, and

using these strategies (i.e., a complex brain) is in place, it can be used for a variety of other purposes including coping with environmental, ecological, technological, linguistic, and other challenges. (Sergey Gavrilets and Aaron Vose, 'The Dynamics of Machiavellian Intelligence', *Proceedings of the National Academy of Science* [U.S.] 103 [2006]: 16823–28.

132 ***Opinions differ as to which of the two has driven the huge increase in brain size*** According to Robin Dunbar and Susanne Shultz in their article 'Evolution in the Social Brain', it "may have been the particular demands of the more intense forms of pair-bonding that was the critical factor that triggered this evolutionary development" (*Science* 317 [2007]: 1344–47).

132 ***Evolution strongly favours strategies that minimise the risk of loss*** See, for example, Michihiro Kandori, George Mailath, and Rafael Rob, 'Learning, Mutation and Long Run Equilibria in Games', *Econometrica* 61 (1993): 29–56.

132 ***'it is far better to earn the confidence of the people than to rely on [force]'*** W. K. Marriott, translator's preface to Nicolo Machiavelli, *The Prince*, Project Gutenberg, www.gutenberg.org/etext/1232. Machiavelli's own words are that 'it is necessary for a prince to have the people friendly, otherwise he has no security in adversity', and this advice is repeated throughout his essay. It is advice that he had already offered in an earlier book, *The Discourses on the First Ten Books of Titus Livius*, in which he says that the best remedy for popular hostility is 'to try to secure the goodwill of the people' (www.gutenberg.org/etext/10827). Perhaps this sage advice from 350 years ago could be heeded by some modern rulers.

The historian Mary Deitz claims that The Prince is actually a 'political act', 'an act of deception', a piece of 'duplicitous advice' designed to restore a republic in Florence by tricking a 'gullible and vainglorious prince', Lorenzo de Medici, into implementing policies that would 'jeopardise his power and bring his demise'. For this and counter-arguments, see John Langton and Mary G. Deitz, 'Machiavelli's Paradox: Trapping or Teaching the Prince', *American Political Science Review* 81 (1987): 1277–88.

133   *misplaced trust can even be a factor in extinction* The dodo, for example, was easy prey because it did not fear humans. It was also vulnerable, however, to predators such as cats and rats, as well as to habitat destruction. The last one probably died around 1690 (D. L. Roberts and A. R. Solow, 'When Did the Dodo Become Extinct?' *Nature* 426 [2003]: 245).

According to some reports, cooked dodo tasted terrible.

133   *mistrust should always predominate* Brian Skyrms (Presidential Address to the Pacific Division of the American Philosophical Association, March 2001, www.lps.uci.edu/home/fac-staff/faculty/skyrms/StagHunt.pdf) describes what happens this way: 'If mutations are unlikely and the probability of mutations is independent across individuals, the probability of mutations taking you from the basin of attraction of the cooperative equilibrium to that of the non-cooperative one is much larger than the probability of mutations taking you in the opposite direction. Therefore the population spends most of its time not cooperating.'

He states that 'perhaps, someone might say, we are evolved to be the kind of species with a predilection for cooperation – with some initial but defeasible predilection for trust in cooperative enterprises built into our nature. The same problem now emerges in even grander evolutionary terms. Should we expect evolutionary dynamics to respect pay off dominance when it conflicts with risk dominance? The answer usually delivered up by contemporary evolutionary game theory is "No." In the long run, one should expect to see the risk-dominant equilibrium almost all the time' ('Trust, Risk and the Social Contract', Synthese 160 [2008]: 21–25).

134   *evolutionarily stable strategy* This self-explanatory term was introduced by J. Maynard Smith and G. R. Price in 'The Logic of Animal Conflicts', *Nature* 246 (1973): 15–18.

134   *arranged for a well-known TV presenter to record two TV clips* See Richard Wiseman, *Quirkology: The Curious Science of Everyday Lives* (New York: Basic Books, 2007). People who simply listened to the words without seeing the video did much better than those who relied on visual cues; three-quarters of them got the answer right. So perhaps sight is more deceptive than sound.

135 **the oldest confidence trick of all** The term *confidence man* was coined in 1849 by the *New York Herald* to describe the deceptions of William Thompson (Johannes Dietrich Bergmann, 'The Original Confidence Man', *American Quarterly* 21 [1969]: 560–77).

Confidence tricks have formed the basis of many film plots. *The Sting* is probably the best known, but David Mamet's 1987 directorial debut film, *House of Games*, takes some beating for its con within a con within a con, and also for its publicity tag line, 'human nature is a sucker bet'.

136 **credible commitment** This can be a matter of belief as much as fact. When I was a visitor at a Cambridge college in the United Kingdom, I came across a story that encapsulated this important point. The story concerned an argument that took place in the Victorian era between a college master and the college chaplain. A discussion had developed over port and walnuts as to whether priests or judges had more power. The chaplain argued that priests have more because 'a judge can only say, "You are going to be hanged," but a priest can say, "You are going to be damned".' 'Ah,' cackled the unbelieving master, 'but when a judge says, "You are going to be hanged" you *are* hanged.'

So far as the unbeliever was concerned, only the hanging threat was credible. To some believers, though, both threats would be equally credible. Their power in any case is reflected in Blaise Pascal's slightly tongue-in-cheek pragmatic reason for belief in the existence of God. Put briefly, if you believe in God and he does exist, your reward in heaven is infinite, while if he doesn't exist, you haven't lost much. If, however, you *don't* believe in God and he *does* exist, your loss is infinite if belief is a requirement for entry to Heaven. Your best bet, therefore, whether God exists or not, is to believe that he does.

The argument can be presented in terms of game theory: Your strategy is either to believe or not to believe. The decision matrix of rewards or losses is:

	God Exists	God Does Not Exist	Total Gain from Belief Strategy
Believe	∞	Cost of believing	∞
Don't Believe	$-\infty$	Save the cost of believing	$-\infty$

Pascal's wager is more than a tricky point in philosophy. It is an argument that is used by many Christians today to justify or bolster their faith, and to persuade others to join them in that faith, with the threat of infinite loss if they don't. Interestingly, exactly the same argument can be advanced for several other religions, which leaves one wondering which to choose, if any.

137 *Game theorists offer two basic ways . . . to demonstrate credible commitment* These approaches are summarised, with many excellent examples from the business and commercial worlds, in Avinash Dixit and Barry Nalebuff, *Thinking Strategically: The Competitive Edge in Business, Politics, and Everyday Life* (New York: Norton, 1991), 144–61. I am also considerably indebted to Professor Bob Marks from the Australian Graduate School of Management for many helpful discussions, not to mention providing me with a copy of his lecture notes on the subject.

137 *actors . . . unlikely to be offered any more roles if they don't turn up for each performance* This is exactly what happened to the British actor Stephen Fry when he famously failed to show up one night for his role in the West End play *Cell Mates* in 1995. It was a result of Fry's bipolar disorder, which had suddenly caused severe depression and an inclination to suicide. Fry has since spoken publicly about his disorder in a wonderful two-part TV series, *Stephen Fry: The Secret Life of a Manic-Depressive*, which was first broadcast on BBC Two in September 2006.

140 *'bringing us to the edge of the nuclear abyss'* Adlai Stevenson, speech. Reported in *New York Times*, February 26, 1956.

140 *Burn your bridges* The ultimate in burning your bridges comes from some primitive fungi and algae that gave up their individuality to form a combined organism called a 'lichen' (with the algae harvesting light for energy and the fungi extracting chemical nutrients from the environment). The original fungi and algae have long since become extinct, but lichens continue to thrive. Some bacteria showed a similar commitment early in cellular evolution when they chose to use living cells as homes in return for providing the cells with energy. Those bacteria eventually lost their ability to exist independently, becoming the mitochondria that we harbour today.

This is not to suggest that any of these species made a conscious decision to limit their own options by burning their bridges. Evolutionary pressures took care of that. The only species capable of voluntarily limiting its own options is Homo sapiens, and we need to learn how to do it more effectively sooner rather than later. If we don't, nature is likely to do it for us.

141 *'exquisite carnal comedy'* Review by Gunnar Bjornstrand (www. lovefilm.com).

142 *Generosity . . . a subset of altruism* Paul Zak, Angela Stanton, and Sheila Ahmadi, 'Oxytocin Increases Generosity in Humans', PLoS ONE 11 (2007), e1,128.

PLoS ONE is an open-access Internet journal. By publishing in it, the authors showed both altruism and generosity.

142 *Harry Lauder* (later Sir Harry Lauder) wrote songs that included 'Roamin' in the Gloamin'' and 'I Love a Lassie.' He made some twenty-two trips to the United States, and was not as mean as he made himself out to be. During the First World War he led fundraising efforts for war charities and voluntarily entertained troops in France under enemy fire.

143 *feeling . . . may be related to the concentration of oxytocin* Paul Zak, Angela Stanton, and Sheila Ahmadi, 'Oxytocin Increases Generosity in Humans', *PLoS ONE* 11 (2007), e1,128.

143 *charitable giving activates reward regions in the brain* J. Moll, F. Krueger, R. Zahn, M. Pardini, R. de Oliveira-Souza, and J. Grafman, 'Human Fronto-Mesolimbic Networks Guide Decisions About Charitable Donation', *Proceedings of the National Academy of Sciences* (U.S.) 103 (2006): 15623–28; D. Tankersley, C. J. Stowe, and S. A. Huettel, 'Altruism Is Associated with an Increased Neural Response to Agency', *Nature Neuroscience* 10 (2007): 150–51; W. T. Harbaugh, U. Mayr, and D. R. Burghart, 'Neural Responses to Taxation and Voluntary Giving Reveal Motives for Charitable Donations', *Science* 316 (2007): 1622–25.

According to Harbaugh, Mayr, and Burghart,

civil societies function because people pay taxes and make charitable contributions to provide public goods. One possible motive for charitable contributions, called 'pure altruism', is satisfied by increases in the public good no matter the source or intent. Another possible motive, 'warm glow', is only fulfilled by an individual's own voluntary donations. Consistent with pure altruism, we find that even mandatory, tax-like transfers to a charity elicit neural activity in areas linked to reward processing. Moreover, neural responses to the charity's financial gains predict voluntary giving. However, consistent with warm glow, neural activity further increases when people make transfers voluntarily. Both pure-altruism and warm-glow motives appear to determine the hedonic consequences of financial transfers to the public good.

The warm glow has its limits, though. When people in Atlanta, Georgia, were questioned about their willingness to pay for pond meshing to save migratory waterfowl from oil pollution, one group was told that their donation would save 2,000 birds. The figure was raised to 20,000 birds for a second group, and 200,000 for a third group. Their willingness to pay was the same in each case! (Peter Diamond and Jerry A. Hausman, 'On Contingent Valuation Measurement of Non-Use Values', in *Contingent Valuation: A Critical Assessment*, edited by Jerry A. Hausman [Amsterdam: Elsevier, 1993], 24).

143 *Charities recognise this response* A British television advertisement for the charity World Vision said, 'You can sponsor a child for just 16p a day, but what you feel in return is priceless.' It worked; both my wife and I now sponsor children, and what we feel in return is indeed priceless.

144 *All he had . . . done was use an indelible marker* This was the notorious Summerlin case (see Joseph Hixson, *The Patchwork Mouse* [London: Archer Press, 1976]), which is not to be confused with the Arizona murder case of the same name. Only the second resulted in the death penalty (http://caselaw.lp.findlaw.com/cgi-bin/getcase.pl?court=US&navby=case&vol=000&invol=03-526).

In another instance of scientific fraud, a senior paleontologist claimed to have found fossils in unlikely locations. Despite the personal trauma it involved, and the damage to another person's career, the professor with whom I was studying the subject at the time eventually published an article in which he said that these claims just had to be fraudulent. (It turned out that the man had simply bought the fossils from a shop.) In game theory terms, my professor's personal stress in performing the exposure was more than offset by the loss that he would otherwise have felt from the damage to truth and honesty in his field if he had not exposed the fraudster.

The scientist concerned was V. J. Gupta. His misdemeanours were exposed by John Talent (J. A. Talent, 'The Case of the Peripatetic Fossils', *Nature* 338 [1989]: 613–15). John later wrote an extended account of the affair, 'Chaos with Conodonts and Other Fossil Biota: V. J. Gupta's Career in Academic Fraud: Bibliographies and a Short Biography', *Courier Forschungsinstitut Senckenberg* 182 (1995): 523–51.

144 **'Networks based on reciprocal moral obligation'** Francis Fukuyama, *Trust* (New York: The Free Press, 1995), 205.

144 **mateship** Macquarie Dictionary (Melbourne: Palgrave Macmillan, 2006). Mateship may have had its source in Australia's convict origins. It is certainly associated with a mutual dislike of authority figures. One of our family stories is that this is why grandparents and grandchildren often get on so well – they have a common enemy.

144 **Bowling Alone** Putnam argues that the participation of individuals in American communities has dropped off drastically over the last forty years. 'We are bowling alone and not in leagues. We are voting at lower rates. We belong to fewer clubs and participate in those we do belong to at lower rates. We are less likely to participate in organized religion. We are joining unions and professional organizations at lower rates. We are spending less time socializing. We donate less to charity (as a percentage of income). We trust our neighbors less. More of us are lawyers' (Robert Putnam, *Bowling Alone: The Collapse and Revival of American Community* [New York: Simon & Schuster, 2000]).

Putnam summarised his analysis of the year 2000 Social Capital Community Benchmark Survey in his 2007 Johan Skytte Prize lecture 'E Pluribus Unum: Diversity and Community in the Twenty-first Century', *Scandinavian Political Studies* 30 (2007): 137–74.

146 **Seeing themselves as members of the same village community** The advantages of belonging to a group have been lumped together by sociologists under the heading 'social capital'. One of

those advantages is trust. The validity of the concept of social capital has been extensively analysed by Joel Sobel in his essay 'Can We Trust Social Capital?' *Journal of Economic Literature* 40 (2002): 139–54.

146 ***ethnic, cultural, and religious differences have been major sources of conflict*** Readers will be aware of many historical and contemporary examples. One of the worst in modern times is the situation in Africa, where ethnic differences have been responsible for many civil wars. Africa was subdivided by Western powers in the nineteenth century into 'countries' whose boundaries cut right across traditional tribal boundaries. The mistrust among different tribes that have been so circumscribed has been reflected in wars of genocidal proportions in Rwanda, Sudan, Kenya, Chad, and many other areas.

146 ***Influential thinkers . . . argued that world government was the only way*** For example, Bertrand Russell in *Has Man a Future?* (London: Penguin, 1961). A good summary of the history of world government is given at http://en.wikipedia.org/wiki/World_government.

147 ***United Nations Charter*** www.un.org/aboutun/charter.

147 ***ritual*** Animals have many rituals, for example, in the form of displays and other behaviour patterns, and these demonstrate credible commitment because the animals' brains are hardwired to stick to that commitment.

Human rituals are different. Some scientists even think that we ought not to use the word *ritual* to describe this behaviour, and that a word like *habituation* might be better. The difference lies in the fact that we have a choice; no matter how deeply buried are the emotional, psychological, and biological origins of the ritual, our brains are still wired in such a way that we can choose how to react to it. See, for example, 'A Theoretical Framework for Studying Ritual and Myth' (Emory Center for Myth and Ritual in American Life http://www.marial.emory.edu/research/theoretical.html]).

147 ***'a little known ritualised fishing ceremony'*** David Attenborough, *Life on Air: Memoirs of a Broadcaster* (Princeton: Princeton University Press, 2002), 133.

148 **James George Frazer** Frazer is best known as the author of The Golden Bough, a wide-ranging comparative study of mythology and religion that very much shocked its audience when it was first published in 1890, because it presented the Christian image of the Lamb of God as a relic of a pagan religion. Frazer may have been forced to back off on this interpretation by the time that he published a third edition, where among other changes he removed his analysis of the crucifixion to a speculative appendix.

The title comes from the Greek myth that is retold in Virgil's Aeneid, in which Aeneas journeys to Hades with the Sybil and presents the golden bough to the gatekeeper in order to gain admission.

148 **Wittgenstein . . . argued that Frazer had ignored the expressive and symbolic dimensions** A summary and discussion of this disagreement has been given by Jacques Bouveresse ('Wittgenstein's Critique of Frazer', *Ratio* 20 [2007]: 357–76). According to Bouveresse, Wittgenstein also argued it was 'an error to try to explain the powerful emotions evoked even today by traditions such as fire festivals (which may once have involved human sacrifice) by searching for their causal origins in history or prehistory'.

148 **a ritual handshake, a binding contract** Specifically, in Scotland when it comes to agreeing on the sale of a house. See, for example, www.georgesons.co.uk/offer.html.

149 **Relationship counsellors place strong emphasis on the role of trust** For example, see Richard Nelson-Jones, *Human Relationship Skills*, 2nd ed. (Sydney: Holt, Rienhart & Winston, 1991), 141–42.

149 **'the motivating efficacy of manifest reliance'** Philip Pettit, 'The Cunning of Trust', *Philosophy and Public Affairs* 24, no. 3 (Summer 1995): 208.

149 **'the trust mechanism'** Daniel M. Hausman, *Trust in Game Theory*, unpublished paper, 1997, philosophy.wisc.edu/hausman/papers/trust.htm, used with permission.

150 **Reader's Digest *conducted an experiment*** Australian *Reader's Digest*, August 2007, 36–43.

151 **'Trust materializes reliably'** Philip Pettit, 'The Cunning of Trust', *Philosophy and Public Affairs* 24, no. 3 (Summer 1995): 202.

152 **person-centred approach** This approach, in which the client is offered 'unconditional positive regard', is now used by many professional counsellors. It was pioneered by Rogers and described in his groundbreaking book *Client-Centered Therapy: Its Current Practice, Implications, and Theory* (Boston: Houghton Mifflin, 1951). For more information, see www.carlrogers.info/.

152 **BookCrossing** www.BookCrossing.com.

Chapter 7

155 **Mrs. Doasyouwouldbedoneby and Mrs. Bedonebyasyoudid**
When the five-year-old Julian Huxley (later to become famous as a zoologist and founder of UNESCO) read *The Water Babies*, he asked his grandfather (the redoubtable T. H. Huxley, known as 'Darwin's bulldog') whether he had ever seen a water baby. His grandfather's reply was a masterpiece of adult writing to a child without patronising:

> My dear Julian,
>
> I never could make sure about that water baby. I have seen Babies in water and Babies in bottles; but the Baby in the water was not in the bottle and the Baby in the bottle was not in the water. My friend who wrote the story of the Water Baby was a very kind man and very clever. Perhaps he thought I could see as much in the water as he did. There are some people who see a great deal and some who see very little in the same things.
>
> When you grow up I dare say you will be one of the great-deal seers and see things more wonderful than Water Babies where other folks can see nothing. (Julian Huxley, Memories [London: Allen & Unwin, 1970], 24–25.)

155 **repeated interactions are an important key to finding cooperative solutions** This is on the assumption that the sequence of interactions is indefinite – in other words, its end cannot be predicted. If it does have a definite, predictable end, then the Prisoner's Dilemma and other social dilemmas still continue to exert their stranglehold, at least in theory, because we can always look forward and reason backwards to come up with the conclusion that it is rational to defect on cooperation at the last step (the end game), and thence on the step before that, and the step

before that, and the . . . By reasoning from the end backwards, when we know that there is a definite end, the whole game unravels.

156 **Mrs. D and Mrs. B represent two extreme approaches** The difference between them is encapsulated in the story given about the relative power of judges and priests (see p. 234). Mrs. B represents the immediate, sure retribution offered by the judge. Mrs. D offers the remote, potential benefits and punishments that will follow if you believe the priest.

156 **ethic of reciprocity** While I was writing this chapter, I came across a wonderful example from the show *Dr. Phil*. Talking about mass killers, Dr. Phil made the point that the one thing they had in common was that they had all felt excluded. 'What would have been the effect,' he mused, 'if someone had said at some stage, "hey, come and sit with us."'

156 **the ethic of reciprocity . . . is advocated by of most of the world's major religions** A listing of citations of religious writings that express the ethic of reciprocity can be found at tralvex.com/pub/spiritual/index.htm#GR. The traditions included (as the compilers describe them) are ancient Egyptian, Baha'i, Buddhism, Christianity, Confucianism, Hinduism, Humanism, Native American spirituality, Islam, Jainism, Judaism, Shinto, Sikhism, Sufism, Taoism, Unitarianism, Wicca, Yoruba, and Zoroastrianism.

157 **Jesus propounded it in the Sermon on the Mount** The Bible reports Jesus as giving the same message in two separate sermons (which may or may not have been the same sermon, and which in both cases may simply be summaries of Jesus' teaching by the two authors concerned). Matthew, describing the Sermon on the Mount (Matthew 5–7), reports Jesus as saying, 'So in everything, do to others what you would have them do to you' (7:12, NIV). Luke (6:20–49) describes the 'Sermon on the Plain', in which Jesus is reported as saying, 'Do to others as you would have them do to you' (Luke 6:31, NIV).

157 **The Prophet Muhammad . . . admonished believers to 'hurt no one so that no one may hurt you'** In his last sermon, which has been reported in the hadiths (reports of oral traditions) of many

Islamic scholars, such as hadith 19774 in the Masnud (hadith collection) of Imam Ahmed ibn Hanbal. The text of the sermon (which has several variants) is widely available on the Internet. See, for example, 'The Last Sermon of the Prophet Muhammad', paragraph 2, www.cyberistan.org/islamic/sermon.html.

157 *Confucius said in* **The Analects**, *'Never impose on others what you would not choose for yourself'* The Analects of Confucius, translated by David Hinton (Berkeley, Calif.: Counterpoint, 1998), XV.24.

157 *The Dalai Lama put it in a different . . . form* Quoted in Mabel Chew, Ruth M. Armstrong, and Martin B. Van Der Weyden, 'Can Compassion Survive the 21st Century?' *Medical Journal of Australia* 179 (2003): 569–70.

157 *The ethic of reciprocity is a statement of morality in which many of us believe* The Parliament of the World's Religions, for example, advances the ethic of reciprocity as a basis for world peace and cooperation. This parliament is an ongoing attempt to create a global dialogue of religious faith. Following an initial meeting in Chicago in 1893, and after a lapse of one hundred years, another meeting under the same title was held in Chicago in 1993, followed by one in Cape Town in 1999 and one in Barcelona in 2004. Statistics suggest that their attempts to promote world peace have not yet been noticeably successful. The next meeting is scheduled for Melbourne in 2009.

157 *Pythagoras* Quoted in Sextus Empiricus, *Adversus Mathematicos*, translated by D. L. Blank (New York: Oxford University Press, 1998).

157 *The categorical imperative* Immanuel Kant introduced this concept in his *Foundations of the Metaphysics of Morals*, published in 1785 (available in translation by Lewis White Beck; New Jersey: Prentice Hall [1989]) and continued to expand on it throughout his life.

158 *Mrs. B took a much more sceptical view of human behaviour and human values* It is also the strategy of Friedrich Nietzsche rather than Kant. Nietzsche said, for example, 'The world viewed from inside . . . it would be "will to power" and nothing else' (Beyond Good and Evil, translated by Walter Kaufmann [New York:

Vintage, 1979], section 36). He has also frequently been quoted as saying, 'It is impossible to suffer without making someone pay for it; every complaint already contains revenge', and 'The best weapon against an enemy is another enemy', although these quotes are unsourced.

158 ***American brown-headed cowbird . . . use Mrs. B's retributive tactics in a protection racket*** Jeffrey P. Hoover and Scott K. Robinson, 'Retaliatory Mafia Behavior by a Parasitic Cowbird Favors Host Acceptance of Parasitic Eggs', *Proceedings of the National Academy of Sciences* (U.S.) 104 (2007): 4479–83. The warblers that they prey on are actually 'wild prothonotary warblers', if you must know.

158 ***Rats . . . use Mrs D's*** Claudia Rutte and Michael Taborsky, 'Generalized Reciprocity in Rats', *PLoS Biology* 5 (2007): e196, doi:10.1371/journal.pbi0.0050196. Only female rats were involved in the study, which may or may not be significant.

159 ***Vampire bats will feed blood to others*** G. S. Wilkinson, 'Reciprocal Food Sharing in the Vampire Bat', *Nature* 308 (1984): 181–84.

159 ***Chimpanzees will offer to share meat*** Kevin E. Langergraber, John C. Mitani, and Linda Vigilant, 'The Limited Impact of Kinship on Cooperation in Wild Chimpanzees', *Proceedings of the National Academy of Sciences* (U.S.) 104 (2007): 7786–90.

159 ***will go out of their way to help an unfamiliar human who is struggling to reach a stick*** Felix Warneken, Brian Hare, Alicia P. Melis, Daniel Hanus, and Michael Tomasello, 'Spontaneous Altruism by Chimpanzees and Young Children', *PLoS Biology* 5 (2007): e184. The study also found that young toddlers behave in the same altruistic way!

159 ***reciprocal altruism*** The term was coined by Harvard University biologist Robert Trivers in a 1971 review, 'The Evolution of Reciprocal Altruism' (*Quarterly Review of Biology* 36: 35–57). Trivers listed three prerequisites for reciprocal altruism (i.e., the alternation of donor and recipient roles in repeated altruistic interactions) to occur:

      1. a large benefit to the recipient and a small cost to the donor,

      2. repeated opportunities for cooperative interactions, and

      3. the ability to detect cheats.

160   *'Always forgive your enemies'* Often attributed to Oscar Wilde, but unsourced.

161   *the stolen generation* A particularly poignant example of what happened is depicted in the film *Rabbit-Proof Fence*, which is based on the real story of three young girls of Aboriginal descent who escaped from the orphanage to which they had been taken, and trekked across Australia to rejoin their families.

161   *the present government offered an unconditional apology* This historic apology was delivered by Prime Minister Kevin Rudd on Wednesday, February 14, 2008, in the presence of Aboriginal leaders. The full text is given at en.wikipedia.org/wiki/Stolen_Generations#Apology_text.

162   *the Samaritan paradox* Economist Samuel Bowles makes an interesting case that we may have evolved the capacity to be altruistic because it makes us better at waging war! In 'Group Competition, Reproductive Leveling, and the Evolution of Human Altruism' (*Science* 314 [2006]: 1569–72) he uses a mathematical model to support his argument that when groups are in conflict, altruism helps to protect the group against the costs of conflict. It seems that the sort of mateship that I described in the last chapter, and that provided mutual support for Australian soldiers through two world wars, may be an evolutionarily stable strategy. I wonder what the old Australian bush poets would have made of it all?

162   *'to disarm a Greek, it is only necessary to embrace him'* Lawrence Durrell, *Bitter Lemons* (London: Faber and Faber, 1957), 79.

163   *Prisoner's Dilemma computer tournament* The reward structure in the tournament was in terms of arbitrary points: 3 for mutual cooperation, 1 for mutual defection, while if one player defected while the other cooperated, the defector received 5 and the cooperator received 0. The tournament is described in detail in Robert Axelrod, *The Evolution of Cooperation* (New York: Basic Books, 1984). More technical analyses are to be found in

his original papers: 'Effective Choice in the Prisoner's Dilemma', *Journal of Conflict Resolution* 24 (1980): 3–25, and 'More Effective Choice on the Prisoner's Dilemma', *Journal of Conflict Resolution* 24 (1980): 379–403. Axelrod also published several pioneering papers on its application to the problem of the evolution of co-operation in nature, notably (with William Hamilton) 'The Evolution of Cooperation', Science 211 (1981): 1390–96.

Evolutionary biologist Richard Dawkins wrote in a foreword to the second edition of *The Evolution of Cooperation* that 'the planet would become a better place if everybody studied and understood it'. 'The world's leaders should all be locked up with this book and not released until they have read it,' he said, adding that it 'deserves to replace the Gideon Bible'. Dawkins was indulging in hyperbole, but he had a point. I would certainly like to see *The Evolution of Cooperation* in hotel rooms, along with several other books that provide insights from different angles into how the world works. I would personally include Bertrand Russell's *Sceptical Essays*, Jacob Bronowski's *The Ascent of Man*, David Attenborough's *The Living Planet*, and Simon Singh's *Fermat's Last Theorum* for starters.

164 **'Don't be envious'** Robert Axelrod, *Evolution of Cooperation* (New York: Basic Books, 1984), 110.

165 **the strongest eventually coming out on top** Many people take this to be synonymous with 'survival of the fittest', but this was not what Charles Darwin meant when he coined the phrase.* He meant it to be synonymous with *natural selection* – that is, the survival and propagation of those members of a species that are best fitted to their environment and circumstances. He did not

---

*Darwin used it in *The Origin of Species*, which was published in 1859. In chapter 4, entitled 'Natural Selection: Or the Survival of the Fittest', Darwin wrote that 'this preservation of favourable individual differences and variations, and the destruction of those which are injurious, I have called Natural Selection, or the Survival of the Fittest'. It is often claimed that the British economist Herbert Spencer coined this phrase, and he used it in his *Principles of Biology* (1864, 1:444), yet he credited it to Darwin: 'This *survival of the fittest* [my italics], which I have here sought to express in mechanical terms, is that which Mr. Darwin has called "natural selection," or the preservation of favoured races in the struggle for life'.

necessarily mean 'Nature, red in tooth and claw', although the poet Tennyson took it in this way in his poem 'In Memoriam'.* Darwin meant it to describe *any* situation in which species that are best adapted to their environmental circumstances have the best chance of surviving for sufficiently long to pass their survival characteristics on to their progeny. He would surely have been horrified by the subsequent emergence of its application to human circumstances, and especially its interpretation as the survival of those who can dominate and crush others, which has been used to justify eugenics, racial purification, 'social Darwinism', and ruthless laissez-faire capitalism. The literature and arguments on these topics would surely be sufficient to sink Noah's proverbial ark. The logical fallacies inherent in them have been capably analysed in many places (e.g., see John Wilkins, 'Evolution and Philosophy: Does Evolution Make Might Right?', TalkOrigins Archive (1997), www.talkorigins.org/faqs/evolphil/social.html.

165 *Successful cooperative social groups need their members to be altruistic and cooperative* The first person to recognise the importance of cooperation in evolution was the Russian anarchist Peter Kropotkin, who argued in his 1902 book *Mutual Aid: A Factor of Evolution* (www.gutenberg.org/etext/4341) that 'sociability is as much a law of nature as mutual struggle . . . mutual aid is as much a law of nature as mutual struggle'. Indeed, many experiments have now shown that most animals look after their kin to protect their genetic inheritance not because they know that this is what they are doing, but because those animals that survive and prosper are the ones that have cooperative behaviour encoded in their genes.

---

* The full verse from Tennyson's 'In Memoriam', Canto 56, which refers to humanity, is:

> Who trusted God was love indeed
> And love Creation's final law
> Tho' Nature, red in tooth and claw
> With ravine, shriek'd against his creed

Kropotkin wrote *Mutual Aid* after a journey to eastern Siberia and northern Manchuria. He was clearly looking for a biological justification for socialism, but his observations nevertheless stand as an unbiased account of nature in action. His search was based on a lecture that he had heard at a Russian Congress of Naturalists in January 1880, during which the St. Petersburg zoologist Karl Kessler had spoken on the 'law of mutual aid'. 'Kessler's idea,' Kropotkin wrote, 'was, that besides the law of Mutual Struggle there is in Nature the law of Mutual Aid, which, for the success of the struggle for life, and especially for the progressive evolution of the species, is far more important than the law of mutual contest.'

What he saw during his journey made two lasting impressions. 'One of them,' he said, 'was the extreme severity of the struggle for existence which most species of animals have to carry on against an inclement Nature; the enormous destruction of life which periodically results from natural agencies; and the consequent paucity of life over the vast territory which fell under my observation.' The other was that, 'even in those few spots where animal life teemed in abundance, I failed to find – although I was eagerly looking for it – that bitter struggle for the means of existence, among animals belonging to the same species, which was considered by most Darwinists (though not always by Darwin himself) as the dominant characteristic of struggle for life, and the main factor of evolution.' Instead, he discovered countless examples of 'the importance of the Mutual Aid factor of evolution'.

166 **'kin selection'** Its application to humans was parodied by the pioneering British geneticist J. B. S. Haldane when he replied to a journalist's question by saying: 'Would I lay down my life to save my brother? No, but I would to save two brothers or eight cousins.' Haldane was a very courageous man. When he lay dying of cancer, he wrote a poem that began: 'I wish I had the voice of Homer / to sing of rectal carcinoma' ('Cancer Is a Funny Thing', *New Statesman*, February 21, 1964).

166 **talk on the subject of the Good Samaritan** J. M. Darley and C. D. Batson, 'From Jerusalem to Jericho: A Study of Situational

and Dispositional Variables in Helping Behavior', *Journal of Personality and Social Psychology* 27 (1973): 100–108.

168 *'Give me the child until he is seven . . . and I will show you the man.'* Variously attributed to Ignatius of Loyola (the founder of the Jesuit order), Francis Xavier, or the Spanish Jesuit scholar Baltasar Gracian.

168 *where do such norms come from?* 'The existence of social norms is one of the big unsolved problems in cognitive science. . . . We still know little about how social norms are formed, the forces determining their content, and the cognitive and emotional requirements that enable a species to establish and enforce social norms.' (Ernst Fehr and Urs Fischbacher, 'Social Norms and Human Cooperation', *Trends in Cognitive Sciences* 8 [2004]: 185–90).

168 *Social norms are . . . 'standards of behaviour'* Ernst Fehr and Urs Fischbacher, 'Social Norms and Human Cooperation', *Trends in Cognitive Sciences* 8 (2004): 189. The threat of social disapproval encourages most of us to stick to social norms, although my wife and I do have one family friend who bucks the system by wearing shorts wherever he goes, even to formal occasions. Maybe this wouldn't matter so much in some countries, but this is in middle-class England. One has to admire his courage, especially in the English winter. On the other hand, he doesn't often get invited to formal occasions.

169 *sanctions can range from disapproval to social exclusion and worse* In one particularly nasty case in Australia, hinted at in the autobiography of a policeman who had close links with the community, a whole Aboriginal tribe was murdered because one of its members had raped a white woman (M. O'Sullivan, *Cameos of Crime*, 2nd ed. [Brisbane: Jackson & O'Sullivan, 1947]). Sadly, it is not difficult to come up with parallel cases from just about any country in the world.

169 *sanction* In the best-selling novel *The Eiger Sanction*, by Rodney William Whittaker (written under the pen-name 'Trevanian'), *sanction* is used as a euphemism for *murder*.

169   *people who are suffering from AIDS in Thailand* Seth Mydans, 'Thai AIDS Sufferers Ostracized', International Herald Tribune (*Asia Pacific*), November 26, 2006.

169   *Striking waiters . . . carried concealed cameras* James F. Morton Jr., 'The Waiters' Strike', *New York Times*, June 5, 1912, 10.

169   *a man of Chinese origin who had abandoned his three-year-old daughter* 'Pumpkin's Fugitive Father Nai Yin Xue Captured in the U.S.', News.com.au, February 29, 2008, www.news.com.au/pumpkins-dad-found-guilty/story-0-1225738186918.

170   *A passive observer . . . will [punish] one of the participants who has defected* If the experiment involves the sharing of money, for example, observers have been found to be willing to spend real dollars that they would otherwise have received in order to reduce the transgressor's reward by three times that amount (Jeffrey R. Stevens and Marc D. Hauser, 'Why Be Nice? Psychological Constraints on the Evolution of Co-operation', *Trends in Cognitive Sciences* 8 [2004]: 60–65).

171   **conditional cooperation** Ernst Fehr and Urs Fischbacher, 'Social Norms and Human Cooperation', *Trends in Cognitive Sciences* 8 (2004): 186.

171   *collapse of the social norm* This phenomenon reaches epic proportions at the Bombay railway station, for example, where thousands of commuters regularly risk injury or death by running across the tracks after their train comes into the station, rather than taking the time and making the effort to use the overhead bridge.

171   *combination of psychological ingredients that only we possess* Jeffrey R. Stevens and Marc D. Hauser, 'Why Be Nice? Psychological Constraints on the Evolution of Co-operation', *Trends in Cognitive Sciences* 8 (2004): 60–65. I am not sure that I myself possess all of the qualities listed under all circumstances, especially when it comes to delayed gratification in the presence of chocolate or a good bottle of wine.

172 **Dante's Hell** As described in *The Divine Comedy*, Hell consists of nine levels. Interested readers can find out what punishments they are in for, according to Dante, by taking an online test at www.4degreez.com/misc/dante-inferno-test.mv.

172 **Win-Stay, Lose-Shift does even better than TIT FOR TAT** Martin A. Nowak and Karl Sigmund, 'A Strategy of Win-Stay, Lose-Shift That Outperforms Tit-for-Tat in the Prisoner's Dilemma Game', *Nature* 364 (1993): 56–58.

173 **many variants of the Tit for Tat strategy** For example, there are 'a tit for two tats' (which is weakly dominated by Tit for Tat), 'two tits for a tat" and others. Here I have concentrated on those that seem most relevant to real-life cooperation.

175 **Geographical proximity can produce clusters of cooperators** The pioneers in this field were my old friend and bridge partner Professor (now Lord) Robert M. May of Oxford and Martin A. Nowak ('Evolutionary Games and Spatial Chaos', *Nature* 359 [1992]: 826–29). They maintained their conclusions in the face of an attempted rebuttal (Arijit Mukherji, Vijay Rajan, and James L. Slagle, 'Robustness of Cooperation', Nature 379 [1996]: 125–26) with a robust reply (Martin A. Nowak, Sebastian Bonhoeffer, and Robert M. May, *Nature* 379 [1996]: 126).

178 **hand-washing by men in public bathrooms** Reported to me by a medical scientist who wishes to remain anonymous.

178 **'speak softly and carry a big stick'** According to *Wikipedia*, the phrase was derived from a West African proverb (http://en.wikipedia.org/wiki/Big_Stick_Diplomacy). Teddy Roosevelt used it to describe his threat of military intervention if Colombia failed to support the creation of the nation of Panama in 1903.

178 **'Five Rules for the Evolution of Cooperation'** Martin A. Nowak, *Science* 314 (2006): 1560–63. If you don't read any of the other references that I have given, at least read this! The text can be found at www.hks.harvard.edu/netgov/files/.../02_11_08_seminar_Nowak.pdf.

Chapter 8

181 *One surprising way to produce harmony and cooperation from conflict, disagreement, and discord is to introduce an even more discordant person* Business gurus Barry Nalebuff and Adam Brandenburger provide a market example in their book *Co-opetition* ([London: HarperCollins, 1996], 105–6), saying that it can sometimes be worthwhile for a business to actively encourage competitors – even to pay them to become competitors. They cite the example of Intel, which licensed its original 8086 microprocessor technology to twelve other companies. This created a competitive market for the chip and assured buyers that they wouldn't end up being held hostage by a single supplier. With that guarantee, buyers were willing to commit to Intel's technology.

The opposite effect is also possible, as illustrated by Jaroslav Hašek's *Good Soldier Švek* (translated by Cecil Parrott [London: Penguin, 1974]), a novel set during the First World War, in which the lead character practically brings the German army to its knees by his overenthusiastic cooperation and literal following of orders.

181 **Right Ho, Jeeves** This book was first published in 1922, and it has appeared in many editions since. The quoted story is Jeeves's justification for manoeuvring Bertie into locking everyone out of a country house in the middle of the night and thus becoming universally disliked. His action, though, results in the reconciliation of all parties.

182 *Fader and Hauser* The authors' motivation for their study was not just one of economics. The research was performed at a time when the superpowers appeared to be moving toward some degree of cooperation on nuclear weapons, and the authors wanted to know what might happen to the cooperation if others entered the game.

> One concern is whether the presence of a non-cooperating outside player [such as a rogue nation developing its own nuclear weapons] will encourage or discourage cooperation among the superpowers. . . Consider the dramatic effect of outside players on OPEC. After a decade of highly profitable cooperation (collusion) the cartel collapsed, partially because of increased produc-

tion by non-OPEC nations such as the United Kingdom. Member nations began to cheat more. . . In another example, firms in the U.S. microelectronics industries, adversely affected by the growing influence and economic power of foreign competition, formed [a] Corporation to cooperate on basic and applied research [even though] members risk[ed] loss of competitive research advantage relative to other U.S. firms. (Peter S. Fader and John R. Hauser, 'Implicit Coalitions in a Generalized Prisoner's Dilemma', *Journal of Conflict Resolution* 32 [1988]: 553–82.)

There was concrete evidence, in other words, that non-co-operators can sometimes stimulate cooperation in situations of competition and conflict. The researchers saw the Prisoner's Dilemma as a paradigm for many such situations and so decided to focus on the effect of introducing a non-cooperator to the outcome of the dilemma.

183   **winning program (designed by Australian Bob Marks)** Marks makes the point in a personal communication (May 6, 2008) that, 'of course, when there are more than two players, it is in general not possible to punish one other (say, a defector) without also punishing the third (a possible cooperator). This might be another reason why >?-person games demonstrate a greater degree of cooperation.'

184   **I ostentatiously helped myself to more than my fair share** You will be glad to know that I checked this out with my hosts first. (I *did* want to be invited back!) Their reward (in utils?) was the pleasure of being in on the secret and sharing the experiment with me.

184   **my fellow guests implicitly cooperated** Perhaps there was also a Schelling point involved, in the form of a mutual dislike of people who are too greedy.

185   **Centipede Game** The game and its implications are described in Roger A. McCain, *Game Theory: A Non-technical Introduction to the Analysis of Strategy* (Mason, Ohio: Thomson/South-Western, 2004), 226–31.

186   **Centipede Game does not reflect real-life scenarios** Political scientist Rebecca B. Morton summarises this position in her article 'Why the Centipede Game Experiment Is Important for

Political Science' (in *Positive Changes in Political Science: The Legacy of Richard D. McKelvey's Most Influential Writings*, edited by John H. Aldrich, James E. Alt, and Arthur Lupia [Ann Arbor: University of Michigan Press, 2007]) but goes on to argue that bargaining between politicians in legislatures often follows the pattern of the Centipede Game and can break down for the same reason. The problem can also arise in multistep production processes, such as those required to produce food, distribute it, and sell it (Roger A. McCain, *Game Theory: A Nontechnical Introduction to the Analysis of Strategy* [Mason, Ohio: Thomson/South-Western, 2004], 229). If producers, distributors, and wholesalers each take a cut at their respective stages in the process, there can be precious little left to sell by the time the food reaches the market, as the history of food shortages in some less-developed countries can testify. Similar problems are less likely to occur when there is a bond in the form of an enforceable contract, because the participants would lose out if they did not fulfil their part of the bargain to take the process on to the next stage.

187 **Quantum game theory** The roots of quantum game theory are in papers by mathematicians David Meyer ('Quantum Strategies', Physical Review Letters 82 [1999]: 1052–55) and Jens Eisert (J. Eisert, M. Wilkens, and M. Lewenstein, 'Quantum Games and Quantum Strategies', *Physical Review Letters*, 83 [1999]: 3077–80).

187 **Quantum computers are the computers of the future** For a summary of their prospects, see Deborah Corker, Paul Ellsmore, Firdaus Abdullah, and Ian Howlett, 'Commercial Prospects for Quantum Information Processing', QIP IRC (Quantum Information Processing Interdisciplinary Research Collaboration), December 1, 2005, www.qipirc.org/uploads/file/Commercial%20 Prospects%20for%20QIP%20v1.pdf.

188 **'pseudo-telepathy'** Gilles Brassard, quoted in Mark Buchanan, 'Mind Games', *New Scientist*, December 4, 2004, 32–35.

188 **Quantum strategies . . . improve our chances of cooperation in all of the main social dilemmas except for Stag Hunt** The degree of achievable commitment depends on the degree of en-

tanglement, which need not be 100 per ent for the advantages of quantum game theory over conventional game theory to show through. Adrian Flitney has calculated that it needs 62 per cent entanglement for quantum strategies to yield an advantage over normal strategies in Chicken. Only 46 per cent entanglement is needed for quantum strategies to yield an advantage in the Prisoner's Dilemma. In the Battle of the Sexes any degree of entanglement helps. The main disappointment is Stag Hunt, in which it has turned out that quantum strategies offer no particular advantage (Adrian P. Flitney, Ph.D. thesis, University of Adelaide, 2005, and A. P. Flitney and D. Abbott, 'Advantage of a Quantum Player Against a Classical One in 2×2 Quantum Games', *Proceedings of the Royal Society* [London] A 459 [2003]: 2463–74).

189   *the 'Einstein-Podolsky-Rosen paradox'* A. Einstein, B. Podolsky, and N. Rosen, 'Can Quantum-Mechanical Description of Physical Reality Be Considered Complete?', *Physics Reviews* 47 (1935): 777.

190   **entanglement means that there can be no Nash equilibria in pure strategies** J. Eisert and M. Wilkins, 'Quantum Games', *Journal of Modern Optics* 47 (2000): 2543–56. See also S. C. Benjamin and P. M. Hayden, 'Comment on "Quantum Games and Quantum Strategies",' Physical Review Letters 87 (2001), arxiv. org/abs/quant-ph/0003036.

Eisert and Wilkins laid down a prescription for the development of quantum games from their 'classical' version, although Benjamin and Hayden showed that some of their specific conclusions were wrong. Their prescription for a specific quantum strategy to solve the two-person Prisoner's Dilemma, for example, doesn't work (a completely random strategy is best), although it is possible to develop a specific 'best' strategy for Prisoner's Dilemmas that involve more than two people.

Game theorists Bob Marks and Adrian Flitney have separately commented to me that a Nash equilibrium *may* actually be available in coordination games such as the Battle of the Sexes. Flitney goes on to say that this 'does not change the general idea that in two-player quantum games every strategy will have a counter strategy'.

Index

About the Author

LEN FISHER, PH.D., is the author of *Weighing the Soul* and *How to Dunk a Doughnut*, which was named Best Popular Science Book of the year by the American Institute of Physics. He has been featured in the *Wall Street Journal* and the *San Francisco Chronicle*, and on the BBC, CBS, and the Discovery Channel. He is Visiting Research Fellow in Physics at the University of Bristol and is the recipient of an IgNobel Prize for calculating the optimal way to dunk a doughnut. He divides his time between Wiltshire, England, and Blackheath, Australia.

Notes

Notes

Notes

Notes

Notes

Notes

Notes

Notes

Notes

Notes